Introducing Mineralogy

Other Titles in this Series:

For further details of these and other Dunedin Earth and Environmental Sciences titles see
www.dunedinacademicpress.co.uk

INTRODUCING
MINERALOGY

John Mason

EDINBURGH ◆ LONDON

Published by
Dunedin Academic Press Ltd
Hudson House
8 Albany Street
Edinburgh EH1 3QB

London office:
352 Cromwell Tower
Barbican
London EC2Y 8NB

www.dunedinacademicpress.co.uk

ISBNs
9781780460284 (Paperback)
9781780465203 (ePub)
9781780465210 (Kindle)

British Library Cataloguing in Publication data
A catalogue record for this book is available from the British Library

Typeset by Makar Publishing Production, Edinburgh, Scotland
Printed in Poland by Hussar Books

Contents

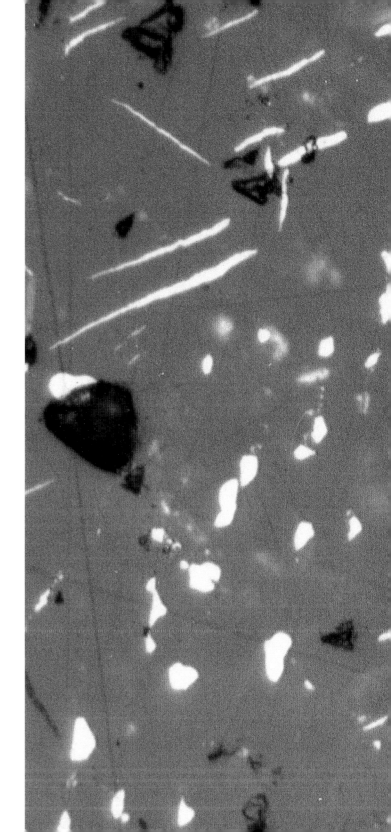

Acknowledgements

The great majority of the illustrations have been provided by the author. The image used for Fig. 1.9 is © the National Museum of Wales, Cardiff, from where Dr Jana Horak is thanked for her help. Likewise Kurt Hollocher of Union College, New York State, USA is thanked for permission to use the photomicrographs in Fig.s 5.3A, 5.3B, 5.4 and 5.6A. Tom Cotterell is thanked for the photographs used in Fig. 4.2. The stibnite image at Fig. 4.10 has been reproduced by permission of Dr. Robert Lavinsky, The Arkenstone, USA. Fig.s 1.16 and 4.9 are reproduced by permission of the Natural History Museum, London. Dr David Green is thanked for the image used in Fig. 5.7A and Andy Tindle is thanked for Fig. 5.7B. Table 1.2 (the Periodic table) is reproduced by permission of Shutterstock © charobnica.

Note: all terms initially highlighted in **bold** are defined in the Glossary at the end of the book.

List of illustrations and tables

Title page illustration

A Star Sapphire: Asterism, which results in starburst-like patterns within crystals, is caused by reflections off internal inclusions of other minerals: when well developed in gem-quality blue corundum, the stone is termed a star-sapphire.

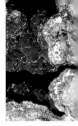

Prologue: a mineral prospector's tale

Pulling onto the side of the road, the air was thick with a single smell – that of splintered conifers. I locked the Land Rover and scrambled down to the river, now running at normal levels again. The flash-floods that had struck North Wales two nights before had been of gargantuan proportions, with many old bridges swept clean away; trees carried out into Cardigan Bay were so numerous that for a time they were declared a hazard to shipping. I had panned this river for gold from time to time and was sure that there was a chance of finding something interesting following such an event.

Walking along the river-bed, I was astounded at how great sections of the banks had simply disappeared. High on either side was a tidemark of branches and other debris left by the floodwaters. Reaching a newly-exposed section of bedrock, one of several that I was heading towards, I stopped to examine it. Within moments I had prised three flakes of gold from the cracks into which they had lodged. I now knew I was the first one here since the storm. Excitement mounted; the hunt was on.

I continued on downstream, checking other areas of rock with more flakes appearing. Then finally, close to the edge of a waterfall, something shining up at me stopped me in my tracks: the edge of a much larger nugget jammed in a crack, revealed where the raging waters had scoured away the thick moss that used to cover this damp outcrop. Whether it had lain there for years undetected, or whether the force of the flood had picked it up, carrying it along until it dropped fortuitously into the crack, remains unknown to this day. I knelt down and carefully, oh so carefully, prised it from its lodgings: six and a half grams of rich yellow gold, beautifully water-worn. It dropped with a loud 'plunk' into my film canister. I put the lid on firmly, buttoned up my shirt pocket and headed for home. Mineralogy, I mused to myself, certainly has its moments.

1   The basics of mineralogy

1.1 What is a mineral?

'In general terms, a mineral is an element or chemical compound that is normally crystalline and that has been formed as a result of geological processes' – E.H. Nickel, writing on behalf of the International Mineralogical Association in the peer-reviewed journal *Canadian Mineralogist* in 1995.

Why use the caveat 'general'? Well, there are just a few exceptions, even though virtually all minerals satisfy these criteria. One interesting exception is mercury. This familiar metal occurs in its elemental or native form in certain ore-deposits: its melting-point is −38.8 degrees Celsius, so that in most parts of the world it would occur in the liquid, rather than crystalline, state. However, it has formed as a result of geological processes and is regarded as a mineral. Likewise, there are a small number of naturally formed compounds that are amorphous, meaning that they are never found in a crystalline state: the copper carbonate georgeite is an example of an amorphous compound that is officially regarded as a mineral.

These few exceptions apart, we have a very straightforward definition to work with. In the same paper, the author goes on to make the distinction that compounds that have formed on materials of anthropogenic origin, such as metal-rich smelter-slags, are not to be regarded as minerals. Compounds that have formed on substrates of natural origin that have been exposed to geological processes by anthropogenic activities – such as tunneling or making road-cuttings, which could result in rocks being exposed to weathering agents – can be regarded as minerals.

At the time that this definition was published, there was a degree of dissent within the mineral collecting community, because the compounds occurring as a result of weathering of ancient smelter-slags often form beautiful, colourful crystals lining cavities in the slags. Some extensive collections of 'slag-minerals' have been built up by some collectors – hence their disappointment in reading that such specimens were no longer regarded as minerals. One suspects that this aspect of the definition of a mineral, and to what extent human activity should be included or excluded from Nature, will continue to prompt lively debate for years to come!

And what about the crystalline compounds and elements of extraterrestrial origin? Nickel goes on to state that extraterrestrial substances 'were apparently produced by processes similar to those on Earth, and therefore such processes are now called geological, even though the term "geology" originally meant the study of rocks on this planet'. This seems reasonable enough. Terms often have to be adapted as circumstances change through the progress of science. So let us now further dissect the definition, firstly looking at what a chemical compound is and how it forms, and secondly, looking at crystals and crystal formation. We will look at the geological processes that lead to the formation of mineral deposits in later chapters; here we deal with the basics.

1.2 Elements and chemical compounds: a chemistry crash-course

Both elements and chemical compounds consist of the basic units of matter: atoms. An element, in its pure form, consists entirely of atoms of itself: thus pure gold consists entirely of atoms of gold. In nature, such purity is virtually unheard-of, so that the native elements will consist almost entirely of atoms of themselves, but will also include impurities – atoms of other elements. Compounds consist of atoms of two or more elements combined and held together by chemical bonds. So how does that work? To find out, we need to start with the basics: what are atoms, and how they may bond to one another to make cohesive substances?

Any atom consists of a dense central nucleus – making up over 99% of its admittedly tiny mass – which is surrounded by one or more negatively charged electrons. The nucleus consists of one or more protons, which carry a positive charge, and neutrons which, as the name suggests, are neutral, and with a similar mass to protons. The electromagnetic force between the positively charged protons and the negatively charged electrons (opposite charges strongly attract one another) is what binds the whole thing together.

Each chemical element has an **atomic number** which is defined by the number of protons in the nuclei of its constituent atoms. Thus hydrogen – element 1 – has a single proton in its nucleus; copper – element 29 – has 29 protons, and lead – element 82 – has 82 protons, and so on. The number of neutrons plus the number of protons gives a second value - the **mass number**. Because the number of neutrons can vary a little, some elements occur in nature as several varieties with different mass numbers. These varieties are known as **isotopes**. An example is

carbon: it has three isotopes – carbon 12, 13 and 14. An atom of carbon 12 has six protons and six neutrons in its nucleus. It is by far the dominant isotope, comprising 98.89% of carbon in nature. Carbon 13, with six protons and seven neutrons, is stable but rare, making up 1.109% of the total. Carbon 14, also known as radiocarbon, is an unstable, radioactive isotope with six protons and eight neutrons, making up the extremely small remainder.

The electrons that surround the nucleus of an atom exist within a series of orbitals or **shells**, each being a zone of stable energy levels. Working outwards from the nucleus of an atom, the shells are labelled 1–6 and each holds a specific maximum number of electrons.

Shell 1 can hold two electrons, shell 2 can accommodate eight; shell 3 eighteen, shell 4 thirty-two, and so on. For example, hydrogen, consisting of one proton and one electron, only has shell 1. Fluorine, with nine protons and nine electrons, has shells 1 and 2 – so there are two electrons in shell 1 and seven out of a possible eight in shell 2. Neon has ten protons and ten electrons, so that shells 1 (2 electrons) and 2 (8 electrons) are both filled. Sodium, on the other hand, has 11 protons and 11 electrons, with shells 1 and 2 filled with two and eight electrons respectively, but with a single electron in the succeeding shell 3. The configuration of sodium's electrons may thus be written 2,8,1.

The table below shows the electron

Table 1.1 The electron configurations of the first twenty elements

Atomic No	Name	Shell 1	Shell 2	Shell 3	Shell 4	Total
1	Hydrogen	1				1
2	Helium	2				2
3	Lithium	2	1			3
4	Beryllium	2	2			4
5	Boron	2	3			5
6	Carbon	2	4			6
7	Nitrogen	2	5			7
8	Oxygen	2	6			8
9	Fluorine	2	7			9
10	Neon	2	8			10
11	Sodium	2	8	1		11
12	Magnesium	2	8	2		12
13	Aluminium	2	8	3		13
14	Silicon	2	8	4		14
15	Phosphorus	2	8	5		15
16	Sulphur	2	8	6		16
17	Chlorine	2	8	7		17
18	Argon	2	8	8		18
19	Potassium	2	8	8	1	19
20	Calcium	2	8	8	2	20

configuration of elements 1–20, demonstrating how successive shells fill up with increased atomic number:

Electrons may move up or down from one shell to another, but to do this they must either absorb or emit energy: those closest to the nucleus are held, because of their proximity, by the strongest electromagnetic force, so cannot easily be moved outwards. But for those electrons in the outermost shell, things are in most cases somewhat easier. This outermost shell is known as the valence shell: **valence** refers to the number of chemical bonds that an atom can form with other adjacent atoms, and it is controlled by the number of electrons and empty spaces in the valence shell.

A look at the **Periodic Table** of the elements (Table 1.2) shows that the elements are arranged in rows and columns, and that groups of chemically similar elements make up the columns – thus the **halogens** – fluorine, chlorine, bromine, iodine and astatine – occupy column 17 and the **noble gases** – helium, neon, argon, krypton, xenon and radon – make up column 18, and so on. This is no accident; the columns are set out to depict the number of electrons present in the outermost shell. So, the elements in column 1 (the **alkali metals**) have one outermost shell electron, column 2 elements (the **alkaline earth metals**) have two. Jumping momentarily past the **transition metals**, which occupy columns 3–12, column 13 elements have three outermost shell electrons, column 14 elements have four and so on through to column 17 – the halogens, which have seven. In column 18, the outermost shells of atoms of these elements – the noble gases – are full, with eight electrons in each case.

An atom with its outermost shell filled has no ready spaces available for the valence electrons of other atoms. This is important, because to form

Table 1.2 Perodic Table

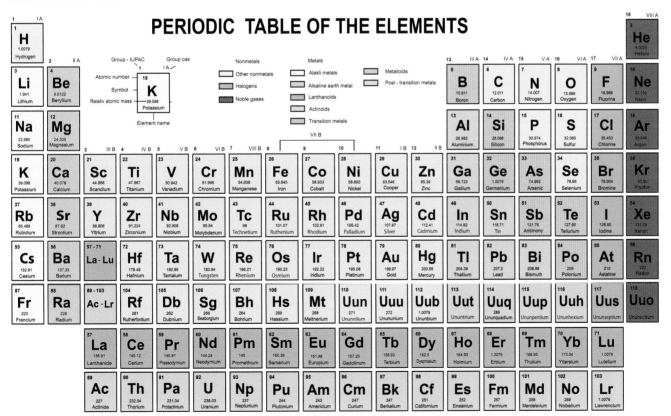

chemical bonds with almost every other element or molecule, atoms not only have to have their own valence electrons available, but also need to have space for those of other atoms. The noble gases, also known as the inert gases, with their 'no vacancies' sign nailed to their full outermost shells, are unreactive in the extreme, forming very few compounds compared to the other elements. Indeed, the term 'stable noble gas configuration' refers to any atom with its outer shell filled.

Contrast the unreactiveness of the noble gases with the eager reaction between a group 1 element and a group 17 element: sodium and chlorine. Sodium has one valency electron and seven vacancies; chlorine has one vacancy and seven valency electrons. The reaction between a piece of sodium metal and chlorine gas is quick and spectacular, as it involves the release of a lot of energy, showing how keen these elements are to bond together and fill those outer shells, in both cases attaining that stable noble gas configuration. The result? Sodium chloride – the common mineral halite, which is also an industrial product, salt, in which sodium (2-8-1) has lost an electron to chlorine (2-8-7), giving them the new and stable electron configurations 2-8 and 2-8-8 respectively.

Columns 3–12 – the transition metals – have varying numbers of valence electrons because they are generally able to make available electrons from the next shell down as well. This property means that they can form compounds in which they can be present in a range of oxidation-states. To grasp what that means, we first need to understand how elements form ions,

beginning, again, with the straightforward example of sodium and chlorine.

An atom containing an equal number of protons and electrons, meaning that the negative charges cancel out the positive charges, is electrically neutral – this is known as its '**ground state**'. However, if it goes on to lose one valence electron, it will then have a net positive charge of +1: thus a sodium atom (symbol Na), with its one electron in its outer shell, can easily lose it to become a positively-charged sodium **ion**, symbol Na^+. Conversely, chlorine (Cl), with its single vacancy in its outer shell, will readily gain an electron, resulting in an ion with a net negative charge (Cl^-). Ions are simply atoms, or molecules, that are positively or negatively charged. They can occur in air, in liquids and in solutions and, just as the opposite charges of protons and electrons attract one another inside atoms, positive (cations) and negative (anions) ions will attract one another and bond together to form compounds.

Copper, for instance, is commonly transported in solution as the Cu^{2+} cation. Sulphate molecules in solution occur as the SO_4^{2-} anion. A solution with copper Cu^{2+} cations and sulphate SO_4^{2-} anions will, when evaporated, leave behind the blue crystalline substance copper sulphate hydrate, $CuSO_4.5H2O$, which occurs in nature as the mineral chalcanthite. Note how the Cu^{2+} and the SO_4^{2-} balance one another out to neutralize the charge.

In terms of electron transfer, **oxidation** is the process of electron removal whilst the opposite, **reduction**, involves the gain of electrons. Thus, setting fire to a length of magnesium ribbon oxidizes the magnesium from

Mg to the oxidation state of Mg^{2+} and combines it with oxygen (O^{2-}) to form MgO – magnesium oxide – in which the 2+ and 2- balance each other out to neutralize the charge. Conversely, if you take copper oxide, CuO, and roast it in a furnace with a source of carbon such as charcoal, you will reduce the copper from the oxidation state of Cu^{2+} down to Cu or metallic elemental copper; this type of reaction with a reducing agent is the basis for the **smelting** of many **base-metals**.

Some groups of elements always have a single oxidation state: for example, group 2 elements like magnesium can only be oxidized to the 2+ state. Others, and especially the transition metals, can be oxidized to multiple states: for example, the transition metal vanadium can be oxidized up to V^{5+} and its near neighbour manganese up to Mn^{7+}. The non-transition metal lead forms a lot of Pb^{2+} compounds, including the chief ore mineral, galena, which is lead sulphide, PbS; it also forms Pb^{4+} compounds, but nowhere near as many, as the 4+ oxidation state in lead is relatively unstable compared to the +2 state. A general rule in chemistry is that systems tend to head towards stability, so that the end-product of a chemical process in nature is a stable compound.

1.3 Chemical bonds

When atoms or molecules combine to form compounds, they are said to have bonded together. So what is a bond? Put simply, it is a strong electromagnetic attraction between atoms and ions that allow them to combine together to form substances. There are four key types of bond in naturally-occurring minerals: these are ionic,

covalent, metallic and hydrogen bonds.

Ionic bonds are formed between metals and non-metals: the metal gives its electron(s) to the non-metal as demonstrated above in the case of sodium chloride, and both metal and non-metal exist as ions within the bonded compound, their opposite charges drawing them strongly together.

Covalent bonds are formed between non-metals whose atoms co-operate by sharing outer-shell electrons to obtain a full outer shell, rather than one atom giving an electron and the other receiving it, as in ionic bonding. Covalent bonds can form between atoms of the same element to create molecules, such as H_2 or Cl_2, or between different elements to form molecular ions such as SO_4^{2-} or PO_4^{3-}.

Metallic bonds are what hold metals and their alloys together; in metals, the outer electrons are only weakly held by the nuclei and thus readily part from them, leaving metal cations sitting in a 'soup' of electrons that freely move around – the freedom of movement of the electrons explaining why metals are such good conductors of electricity. The bonding force is manifested as the multiple attractions between the cations and the electrons that surround them.

Hydrogen bonds are important in geology and mineralogy because they affect the properties of water. A single water molecule is asymmetrical, its two hydrogen atoms carrying a slight positive charge: the one oxygen atom, to which the hydrogen atoms are covalently bonded, carries a slight negative charge. Opposites attract, so that the positive hydrogen atoms bond to other negative oxygen atoms. These hydrogen bonds take, for such light molecules,

a lot of energy to break up, explaining the relatively high boiling point of water. This molecular asymmetry makes water a very good weathering agent: it can readily react with ionically-bonded substances. In liquid water, hydrogen bonds are constantly forming and breaking up, with an average of 3.4 other water molecules being bonded to any one water molecule at any given point in time. In the case of ice, the value is 4 and the resulting structure is less dense than that of water. This is abnormal: most substances are at their densest in the solid form. It is also fortuitous; imagine what would happen if, every time water froze, the ice sank.

So some minerals are held together by ionic bonds, like halite (sodium chloride, NaCl); some are held together by a combination of ionic and covalent bonds, such as anhydrite (calcium sulphate, $CaSO_4$) in which the covalent bonds bind the sulphate anion whilst it is joined to the calcium cation by ionic bonding; and the native metals and their alloys feature metallic bonding. Ice consists of covalently-bonded oxygen and hydrogen atoms forming molecules that are then joined by hydrogen bonds.

1.4 Crystallization

Now we know what bonds atoms and molecules to one another, we can start to think about how such particles mass together to form mineral grains – the way in which the constituent atoms and molecules of mineral compounds are packed together. This packing, as it is called, forms a three-dimensional structure – a **crystal lattice**– whose geometrical characteristics are determined by the sizes of the various ions

involved in different compounds.

Let's take the bonding between silicon and oxygen atoms as an example, because variations in how this occurs account for the diversity of a major and important family of minerals – the silicates. The basic silicate anion is a tetrahedron, the silicon atom sat in the middle and four oxygen atoms positioned around it. The silicon–oxygen bonds are covalent in nature and are very strong indeed. The tetrahedra can bond to one another by adjacent silicon atoms sharing one oxygen atom; in the mineral quartz (SiO_2), each silicon atom shares its oxygen atoms with adjacent silicon atoms, resulting in a 3D, **isotropic** framework of covalently-bonded silicon and oxygen (see Fig. 1.1). The physical properties of quartz reflect this strong, equidirectional bonding: it has no preferred direction of breaking (**cleavage**) and it is a hard mineral.

The silicates consist of a variety of arrangements of bonded tetrahedra based on the SiO_4 building-block. Thus the SiO_4 tetrahedra may be isolated with ionic bonds to metallic cations in between them, or they may be joined to form chains or continuous sheets or frameworks. A well-known example of a sheet-silicate is the mica-group mineral, muscovite, in which strongly-bonded polymer-sheets of silicate and aluminate molecules are separated by relatively weakly-bonded potassium ions. As a consequence, muscovite (and any other mica) is strong in one direction – normal to the plane of the sheet structure – so that it will take a hammer-blow, and in thin sheets it can be bent without breaking. In complete contrast, parallel to the plane of the sheet-structure,

Figure 1.1 The structure of quartz
Quartz consists of a three-dimensional
network of SiO$_4$ (silicate) tetrahedra,
each adjacent silicon atom sharing
its oxygen atoms with other silicon
atoms, thereby creating a very strong
covalently-bonded framework.

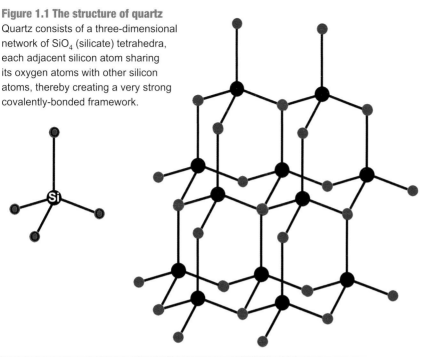

the ionic bonds are so weak and easy
to break that sheets may be easily
peeled apart by hand (Fig. 1.2).

These examples show that the
spatial arrangements of atoms, ions
and molecules in minerals control the
physical properties of their crystals.
The arrangement of these lattice-units
is unique to each mineral species; it
forms an orderly pattern, known as
a **unit cell**, that repeats itself three-
dimensionally (just as the pattern
on a sheet of wallpaper repeats itself
two-dimensionally) to the edges of the
crystal. Within the six **crystal systems**,
there are fourteen different categories
of unit cells, named in reference to their
three-dimensional shape and arrangements of atoms. As an example, we
will look at the most straightforward
category, and one which makes up a
great many mineral species – the cubic
unit cell. It occurs in three varieties:
simple, face-centred and body-centred.

Simple cubic unit cells are as simple
as it gets: each corner of the cube,
known as a **lattice-point**, is marked by
a particle – an atom, ion, or molecule –
depending on what mineral species we
are dealing with. The edges of the unit
cell all connect equivalent lattice-points,
in other words identical particles. Other
particles can be present on the edges
and/or faces of the unit cell, or situated
inside its body. Because a cube has eight

Figure 1.2 Bending mica
Mica, consisting of sheets of strongly-
bonded polymer-sheets of silicate and
aluminate molecules, separated by relatively
weakly bonded potassium ions, peels apart
readily between the sheets, but the sheets
themselves are strong enough that they may
be bent without breaking.

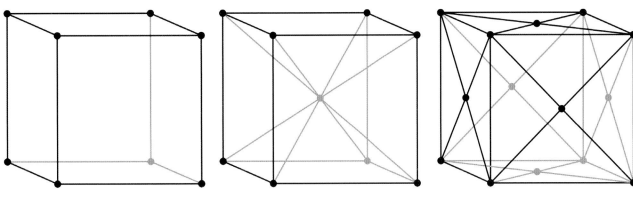

Figure 1.3 Cubic variations
The three basic varieties of the cubic unit-cell – simple, body-centred and face-centred.

corners, a minimum of eight identical particles has to be present to make a simple cubic unit cell.

The **body-centred cubic unit cell** also has those eight identical particles on its corners, but this time there is a ninth identical particle sat right at the centre of the body. The **face-centred cubic unit cell** again has those eight identical particles on the corners of the cube, but in addition has six other identical particles, each occupying the centre of each of the six faces of the cube (Fig. 1.3). This structure clearly has the most identical particles (14 as opposed to 8 and 9 respectively) and is thus the most densely packed out of the three, and is thus termed cubic close-packed. Similar terminologies apply to the other categories of unit cells, so that we see, for example, references to minerals having a hexagonal close-packed structure.

1.5 The crystal systems: describing crystals

Just as unit cells are classified, so are crystals, which are divided into six crystal systems defined by their

morphology, something which is controlled by the ratios of their **crystallographic axes** and which in turn defines their planar and **rotational symmetry** properties. Each **mineral species** crystallizes in one of these six systems: the shape any crystal takes in its outward appearance is known as its **habit**.

The **cubic or isometric** system has all three crystallographic axes of equal length, situated at right angles to each other. Thus equidimensional

(or **equant**) crystals such as cubes or octahedra are formed – these being two of the commonest habits in the cubic system. All crystals belonging to this system have four 3-fold axes of rotational symmetry, each running diagonally from corner to corner through the centre of the cube. They may also have up to three 4-fold axes of rotational symmetry connecting the centres of each of the three opposite pairs of faces (Fig. 1.4).

Figure 1.4 Cubic minerals: pyrite
A 40 mm crystal of pyrite in marl, from Navajun, Spain: a classic example of cubic crystal symmetry.

The **hexagonal** system has three crystallographic axes that intersect at 120° and a fourth – a vertical axis – at right angles to the other three. All crystals have one 6-fold main axis of rotational symmetry and up to six 2-fold axes. There is a subdivision – the **trigonal** division, where the single main axis of rotational symmetry is 3-fold as opposed to 6-fold. Typical crystal habits in the hexagonal system are **prismatic** (elongated) and **bipyramids** (Fig. 1.5).

The **tetragonal** system has three mutually perpendicular axes; two axes are of equal length, while the third vertical axis is of varying length and can be either shorter or longer than the other two. There is a main 4-fold rotational symmetry axis with up to four 2-fold axes; tetragonal minerals can form **tabular** (platy), equant or prismatic habits (Fig. 1.6).

The **orthorhombic** system also has three mutually perpendicular axes, but in this case each has a varying length. There are three main 2-fold axes of rotational symmetry. Orthorhombic minerals typically form tabular to prismatic crystals (Fig. 1.7).

The **monoclinic** system has three axes of varying length, two of which meet at an oblique angle and a third that is perpendicular to the other two. Crystals have a single 2-fold main axis of rotational symmetry; most monoclinic minerals occur in prismatic habits of varying length (Fig. 1.8).

The **triclinic** system has unequal axes, all meeting at oblique angles – in other words, they are asymmetric. Crystals of triclinic minerals tend to be squat or tabular in nature (Fig. 1.9).

Figure 1.5 Hexagonal minerals: vanadinite
Vanadinite crystallizes in the hexagonal system, commonly forming tabular crystals such as these examples, to 10 mm diameter, from Mibladen in Morocco.

Figure 1.6 Tetragonal minerals: wulfenite

Wulfenite commonly crystallizes in the tabular habit, which shows off its tetragonal symmetry, as in this example, with crystals up to 2 mm, from Central Wales. The associated greenish-yellow mineral is pyromorphite.

Figure 1.7 Orthorhombic minerals: barite

Tabular crystals of barite up to 5 cm in height, from the North Pennines of England, showing orthorhombic crystal symmetry.

Figure 1.8 Monoclinic minerals: linarite

Linarite, a secondary basic lead-copper sulphate, crystallizes in the monoclinic system. The sample is from Central Wales and the main crystal is 5 mm long.

Figure 1.9 Triclinic minerals: albite
The asymmetric nature of minerals that crystallize in the triclinic system is illustrated by these twinned albite microcrystals from North Wales.

A further series of terms expresses how well developed a crystal is, from indistinct through to perfectly-formed. The terms are as follows:
anhedral – an irregular grain (Fig. 1.10);
subhedral - an imperfectly-formed crystal (Fig. 1.11);
euhedral - a perfectly-formed crystal (Fig. 1.12).
A mineral can occur in either of these three forms; this depends on the circumstances in which it formed, which we will explore later on. But one general comment will be made: most minerals will tend to form euhedral crystals if at all possible; having the physical room in which to do so is the most important constraint. That is why, in terms of mineral specimens, the best crystals tend to occur lining open cavities (also called pockets or **vugs**).

As well as the crystal habits mentioned in the previous section, there are several others: let's quickly run through them in alphabetical order. **Acicular** crystals are long, thin, needle-shaped prisms. **Bladed** crystals are long and flattened, just like a knife blade. Long, stouter prisms are referred to as **columnar**. **Hemimorphic** crystals are euhedral crystals that can be seen to have two differently-shaped ends. **Hopper-crystals** form when the edges of the crystals grow more quickly than the faces, leaving the latter stepped and concave. **Tetrahedra** consist of crystals with four triangular faces.

However, for the purposes of science, there is a more precise way to describe

Figure 1.10 Anhedral crystals
A hand-specimen showing shapeless (anhedral) grains of dark metallic chromite in a green serpentinite matrix from Shetland.

Figure 1.11 Subhedral crystals
Specular haematite forming imperfect (subhedral), platy 1–2 cm crystals in quartz. Although examination of the specimen reveals some crystal faces, no whole or perfect crystals are present. Sample from North Wales.

crystals: it involves assigning a unique numerical value to each crystal face, a method devised by William Miller, a nineteenth century pioneer. These **Miller Indices**, as they are known, work as follows, and we'll start with a cube again because it is a simple structure.

A perfect cubic crystal has eight corners and six faces. Three crystallographic axes of equal length, two horizontal and one vertical, connect face centres and the centre of the body. These we can call a1, a2 and a3, because they have that equal length, which we'll call x. The point in the dead centre of the body where all three axes meet is known as the **axial cross**. Now, there are three stages to working out the Miller indices for a face on a cubic crystal.

The first thing to notice when you look at a diagram of such a crystal is that each face is only intersected by one axis, at a distance of 1 unit from the axial cross. The other axes go off at right angles out through the other faces and off into the ether. Therefore we'll give these a value of infinity, ∞.

Let's have a go at the top face. It is intersected by the vertical a3 axis but not by the other two, so its parameters may be written ∞,∞,1. Now the critical bit: Miller indices are reciprocal values, so each of these needs to be divided into 1:

1/∞, 1/∞ and 1/1 gives 0, 0 and 1, expressed in brackets thus: {001}.

Each of the three axes intersects two faces, on opposite sides of the crystal: thus, the vertical a3 axis intersects, upwards from the axial cross, the {001} face at the top of the crystal and in the opposite direction – downwards – the face on the bottom of the crystal. For these equal distances but in opposite

Figure 1.12 Euhedral crystals
A perfect (euhedral) 25 mm long doubly terminated crystal of quartz, sat atop well-formed rhombs of ankerite, from the South Wales Coalfield.

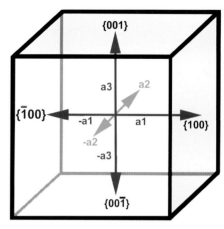

Figure 1.13 Miller-indices
A simple cubic crystal showing the three axes of symmetry, a1, a2 and a3 and the Miller Indices of four of its faces.

directions, the values 1 and –1 are assigned: the negative integer is conventionally written with a bar over the number, hence {00$\bar{1}$} (Fig. 1.13).

Now let's look at an octahedron, another form that can occur in minerals that crystallize in the cubic system. This time, we again have the three axes, a1, a2 and a3, but eight faces, each an equilateral triangle. In this case, every face intersects all three axes at the same distance from the axial cross, so they

are all written {111} – but with bars over numbers where the face is intersected by the negative end of an axis.

These indices are useful not only in specifying individual crystal faces but also in specifying the orientation of cleavage – the planes along which a mineral can be split, as in the example of sheets of mica being peeled apart, described above. Let's take a simple example again, fluorite, CaF_2. This widespread mineral crystallizes in the

cubic system and well-crystallized specimens, forming groups of cubes or more rarely octahedra, can command high prices. But they have to be handled with care: give a fluorite cube a sharp tap and its corners will break off along smooth diagonal planes. So it has a perfect cleavage: with care, a cube may be cleaved at each corner to create an octahedron. The cleavage is termed octahedral and, just as with the crystal faces, the cleavage can be expressed by Miller indices: in this case, {111}. Muscovite mica's cleavage, being one-directional and parallel to the base of a crystal, is written {001} and is also commonly referred to as basal cleavage.

Crystal faces also have names that describe their form. Let's take the tetragonal system as an example. It has two horizontal axes of equal length, a1 and a2, with a vertical c-axis of variable length. A tetragonal crystal thus has a square cross-section parallel to its base, regardless of whether its habit is tabular or prismatic. The top and bottom faces are referred to as **pinacoids**, a term we apply to crystal faces that cut the vertical c axis and are parallel to the horizontal a1 and a2 axes, resulting in a pair of parallel faces with the Miller indices {001} and the negative equivalent {00$\bar{1}$}. The four sides of the crystal that are parallel to the c-axis, which are short if the crystal is tabular or long if it is prismatic, are termed the **prism faces**.

Tetragonal crystals may take the simple form described above or they may be more complex. A frequent form in some tetragonal minerals is the bipyramid, which consists of two four-sided pyramids stuck back-to-back, with eight faces, each of which forms an isosceles triangle oriented along the c-axis. Perfect bipyramids are terminated by sharp points, but combined forms also occur: prism-faces with bipyramidal terminations are one possibility and bipyramids with flat, pinacoidal terminations are another, and all three may occur in combination on the same specimen of the same mineral.

Quite a few minerals can crystallize intergrown as twins. **Twinning** occurs when two separate developing crystals share some of the same crystal lattice points in specific, symmetrical arrangements known as twin laws. In some minerals, characteristic twinning is a major aid to identification. There are several kinds of twinned crystals. Firstly, there are simple twins, which either share a single surface (contact twins) or appear to be passing through one another (penetration twins) (Fig. 1.14). Secondly, there are multiple twins that may be aligned parallel to one another (polysynthetic twins) or not parallel (cyclic twins).

Twinning happens either during or after crystal growth by several mechanisms. Growth-twinning occurs when the lattice of a developing crystal is

Figure 1.14 Crystal-twinning in fluorite
Fluorite forming crystals to 15 mm on edge from the North Pennines, England, exhibiting classic penetration-twinning

disrupted during formation or growth. Some substances are stable in one crystal structure at high temperatures but in another at low temperatures: cooling or heating during growth can destabilize their structure, leading to the development of annealing twins in the interests of stability. Finally, a pre-existing crystal may become twinned during deformation – when earth-movements place rocks under physical strain, leading, for example, to folding. This **strain-related twinning** is also sometimes referred to as glide-twinning.

Under differing conditions, mainly with respect to temperature and pressure, a chemical compound may be more stable when its crystal lattice structure is of one type or another. A classic example is the element carbon: at Earth's surface and in the crust, it tends to occur as the very soft mineral graphite, which crystallizes in the hexagonal system. But at great depth, under the phenomenal pressures found deep in Earth's mantle, the cubic variety – diamond – is the stable one (Fig. 1.15). Many other compounds display this property: in each case each form, crystallizing in a different system, is classified as a separate mineral species. Thus, the compound zinc sulphide (ZnS) occurs as two distinct mineral species: there is the common, cubic form, sphalerite, and the less common, hexagonal form, wurtzite. The two minerals are said to be **polymorphs** of zinc sulphide. Polymorphism is, quite simply, the ability of a solid substance to exist in two or more of the crystal systems.

The suffix 'morph' is also met with in mineralogy when one mineral replaces another, making an exact replica of its crystals. The process can occur in a variety of geochemical environments, but is perhaps most frequently met with in near-surface weathering zones, where slight changes in chemistry can take an existing mineral out of its **stability field**: thus, the copper carbonate azurite ($Cu_3(CO_3)_2(OH)_2$), which is more stable in alkaline environments, is frequently replaced by malachite ($Cu_2CO_3(OH)_2$), which is slightly different in its formula, but more importantly is stable through a greater **pH** range. The azurite crystals are often replaced quite faithfully in terms of their morphology by the fine-grained malachite: such grand imitations are termed **pseudomorphs**. Pseudomorphs of iron oxides after pyrite are another frequently met example: in all cases the term 'after' denotes the pre-existing mineral, if it is possible to discern what that was (Fig. 1.16). A loosely-related

Figure 1.15 Graphite and diamond
Two polymorphs of carbon – a 5 mm euhedral diamond crystal (locality unknown) with, in the background, a lump of graphite from Cumbria, England. In terms of appearance, colour, hardness and other properties they could not be more different!

phenomenon is that of the **epimorph**: an epimorph is formed when one crystalline mineral (mineral 'A') is coated by a layer of another (mineral 'B') but then mineral 'A' is dissolved away. Just as an example, it is not uncommon in mineral deposits for pyrite (FeS_2), which crystallizes in the cubic system and often forms large cubes, to be coated in a layer of small crystals of quartz, which crystallizes in the trigonal system. In weathering-zones of mineral deposits, pyrite readily dissolves away but the chemically resistant quartz remains, so what one ends up with are hollow, cube-shaped boxes made up of quartz: these would be termed quartz epimorphs after pyrite. As with pseudomorphs, a great variety of epimorphs are known to occur: again, identifying the precursor mineral 'A' is not always straightforward (Fig. 1.17).

1.6 Mineral textures

Crystals of the same mineral that occur together – as they almost always do – occur in intergrowths for which there are a number of important textural terms used in their description. Thus, **botryoids** (grapelike shapes) are rounded, multicrystalline growths with a typically smooth surface (Fig. 1.18); a similar texture but more flattened is referred to as **reniform** (kidney-like, Fig. 1.19) and elongate aggregates with a smooth surface are described as **stalactitic**. **Cockscomb** aggregates consist of platy or tabular crystals on edge and crisscrossing (Fig. 1.20). **Dendritic** growths are tree- or fern-like branched masses of crystals. **Drusy** is a textural term describing typically small crystals lining the walls of cavities. **Lamellar** masses are aggregates of flaky minerals such as sheet-silicates. Radiating groups (or **stellate** – star-like) describes multiple crystals all radiating outwards from a common centre (Fig. 1.21), whilst crystals occurring in an oriented criss-cross pattern are said to be **reticulated** (Fig. 1.22). Platy or tabular

Figure 1.16 Pseudomorphs: cassiterite after orthoclase
Euhedral crystals of orthoclase feldspar up to 5–6 cm have been completely replaced by cassiterite. The correct way to describe this specimen is, as with all pseudomorphs, *cassiterite after orthoclase*. Sample is from Cornwall, England. Copyright: the Trustees of the Natural History Museum, London.

Figure 1.17 Epimorphs
Hollow epimorphs of secondary yellow-green pyromorphite (lead chlorophosphate) after 3–5 mm octahedral galena crystals, the remnant shapes of which can still, just about, be discerned. Sample is from Central Wales.

Figure 1.18 Botryoidal texture – malachite
A large hand-specimen of malachite from the Kambove in the Democratic Republic of Congo, showing the botryoidal texture for which the mineral is well known.

Figure 1.19 Reniform texture – hematite
Hand-specimen of haematite (variety kidney-ore), showing the classic reniform texture. Sample from Cumbria, England.

Figure 1.20 Cockscomb texture – barite
Cockcomb barite with multiple crisscrossed platy crystals, forming a small hand-specimen from South Wales.

Figure 1.21 Stellate texture – atacamite
Starburst-like radiating groups of Atacamite crystals have developed on an open joint in granite in this hand-specimen from Copiapo, Chile.

crystals forming sphere-like aggregates are described as **rosettes** (Fig. 1.23).

Poorly or not obviously crystalline minerals again have their own terminology. Minerals forming visible, poorly-formed grains packed together are referred to as **granular** or **massive**, the latter best used where no grain-boundaries can be observed. The term **earthy** refers to crumbly, fine-grained aggregates.

1.7: Mineral properties: cleavage and fracture

Cleavage has already been described in terms of crystallography but there are several descriptive terms that are employed to describe the different types. **Perfect cleavage** is what it says on the tin– the mineral breaking along flat, smooth-sided planes – and the other end of the scale is poor cleavage – which is only discerned with difficulty. In between are imperfect, good, distinct and indistinct (see Figs 1.24–1.26). Some minerals have no cleavage, but all minerals have fracture characteristics, which describe breaks that do not follow cleavage planes: in minerals with good to perfect cleavage this can be difficult to impossible to observe. In minerals with poor or indistinct cleavage the fracture-type can be diagnostic. Quartz has a **conchoidal** fracture, curved like a clam-shell with repeated ridges of a similar nature to those seen in broken glass (Fig. 1.27). Minerals with a similar but less well-developed, curved fracture-pattern are said to have a subconchoidal fracture. Native metals, being malleable to varying extents, do not break so readily, but if ripped apart have break-surfaces with many

Figure 1.22 Reticulated texture – cerussite
Reticulated crystals of cerussite atop an iron-stained mudstone matrix in this hand-specimen from Central Wales.

Figure 1.23 Rosette texture – synchysite
The rare-earth mineral, synchysite, forming rosettes of tabular crystals up to 5 mm on a hand-specimen from North Wales.

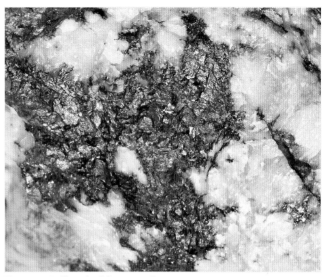

Figure 1.24 Galena – perfect cleavage
Galena, lead sulphide, has a perfect cleavage, with easily visible flat, lustrous cleavage surfaces. Small hand-specimen from Central Wales.

Figure 1.26 Tetrahedrite – no cleavage
No grain is apparent at all in this specimen of fractured tetrahedrite (a complex sulphide of copper, silver, antimony and other metals): it has no cleavage at all. Field of view 40 mm: sample from Central Wales.

Figure 1.27 Conchoidal fracture – flint
Quartz and other forms of silica break with a conchoidal fracture, as well displayed along the sides of this flint implement, where the curved, shell-like embayments are clearly seen.

Figure 1.25 Bournonite – imperfect cleavage
Bournonite is a lead copper antimony sulphide. In this detail of a 30 mm broken part-crystal filling a cavity in quartz, from Central Wales, a grain is visible running NNW–SSE in the image, but it is nowhere near as obvious a cleavage as is displayed by galena. It is thus a poor to imperfect cleavage.

upstanding points: this is referred to as a **hackly** fracture (Fig. 1.28). Some minerals break with no obvious pattern or texture and are said to have an uneven fracture. Some fibrous minerals have a **splintery** fracture, meaning they break up into lots of sharp splinters or fibres (Fig. 1.29).

Figure 1.28 Hackly fracture – copper
A large hand-specimen of intergrown native copper and cuprite from Leicestershire, England, showing the hackly fracture of the copper, with sprigs of the metal sticking up from many parts of the specimen's surface.

Figure 1.29 Splintery fracture – actinolite (below)
Sheaves of fibrous, green crystals of the amphibole group mineral, actinolite, displaying the typical silky lustre and splintery fracture. Specimen from Selåsvatn, Norway: field of view 15 cm in the vertical plane.

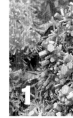

1.8 Mineral properties: colour and lustre

Colour is a self-explanatory term refer-ring to the appearance of a mineral.However, it is not always the most useful of a mineral's diagnostic properties and should only be used in conjunction with the other identi-fication tools. Due to the presence of **trace-elements**, some minerals can occur in a wide range of colours: for example, fluorite is not only purple, it may be green, blue, yellow, pink, white or colourless. However, when finely powdered, it will appear white. This is its **powder colour**, and it is a property that can be helpful in mineral identi-fication. The powder colour, or **streak**, can be examined by drawing a piece of a mineral across the surface of an unglazed tile of white porcelain, also known as a streak-plate. The technique can be diagnostic with some oxides and sulphides, for example – but in the case of many silicates, most of which have a white streak, it is less useful. Fluorite can also appear to change colour when viewed in artificial light then daylight: this is because it can be fluorescent in UV light, including sunlight. A number of minerals exhibit fluorescence, which can be useful in their identification.

Lustre, owing its name to the Latin for light, is a more useful property in terms of identification. It describes how a mineral or crystal surface interacts with, and thus appears in, incident light. A range of terms are employed that may be roughly divided into three categories: **metallic**, non-metallic but bright and non-metallic but dull. It is a subjective parameter with no strict boundaries; the important thing is to only assess it with fresh, unweathered samples.

Weathering can often alter the lustre, dulling it to a non-characteristic form.

Metallic minerals have the appear-ance, especially on fresh crystal faces and cleavage-planes, of polished metal; their surfaces are reflective. Most of the ore minerals fall into this category. A **submetallic** lustre is possessed by some ore minerals that are less reflective.

Minerals whose crystals have a non-metallic but shiny appearance fall into five categories. The most sparkly of the lot have an **adamantine** lustre, diamond being an obvious example. Somewhat less bright, but still noticeably so, are the minerals with a subadamantine lustre. **Vitreous** minerals, as the Latin-derived name suggests, look glassy; a great many of the non-metallic miner-als have this appearance, with quartz a good example. Minerals with a sheet-like or fibrous structure often have a **pearly** or **silky** lustre respectively: the mica-group mineral muscovite (*see* chapter 2) is often described as having a pearly lustre, whilst minerals of the amphibole group of rock-forming minerals often occur as sheaves of fibrous crystals with a silky lustre.

Minerals with low to very low reflectivity are described as **waxy**, **greasy**, **resinous** or just plain **dull**. The first three reflect their visual similarity to the appearances of three familiar substances. Dull minerals include many clays, which do tend to have a lack-lustre appearance.

Some minerals have more unusual optical properties with regard to their interaction with light. **Asterism**, which results in starburst-like patterns within crystals, is caused by reflections off internal **inclusions** of other minerals: when well developed in gem-quality

blue corundum, the stone is termed a star-sapphire. **Aventurescence** is also caused by inclusions of one mineral within another: for example, aventurine is quartz crowded with preferentially-oriented small plates of the green, chromium-bearing mica fuchsite. The reflectance of the fuch-site plates makes good specimens of this form of semi-precious quartz look as though it is full of green glitter.

Chatoyance occurs in minerals which either occur in fibrous bands or contain fibrous bands of other minerals – tiger's eye is a well-known variety of quartz containing bands of fibrous silicates of the amphibole group. The fibres scatter the light so that the bands have a lumi-nous appearance which seems to move as a specimen is turned in the hand.

1.9 Mineral properties: transparency

Transparency refers to the amount of light that can pass through a mineral. A fully **transparent** mineral can be seen through, like a window: a good example is the form of quartz known as rock crystal. However, recrystallization or the presence of inclusions, impuri-ties and structural defects can all affect the transparency: thus quartz is more common in the white form known as milky quartz, through which light will pass, but you cannot see through it: it is **translucent** but not transparent. Very many minerals are translucent, although in some cases the property is limited to the extent that in order to view it a very thin section needs to be cut. Thus, although a lump of the volcanic rock basalt might look grey-black when held in the hand, a **thin section**, cut and then carefully ground down with fine

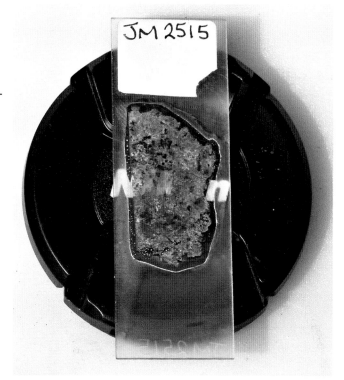

Figure 1.30 Thin section
A thin section of rock mounted on a glass microscope slide: at 30 microns thickness rocks are translucent unless – uncommonly – they are made up of opaque minerals.

Figure 1.31 Polished section
A polished section of galena embedded in a clear, hard resin. This one is rather old and requires repolishing due to scratches. However, because many ore minerals tarnish with exposure to air, periodic repolishing is something that is routinely carried out in any case.

abrasives until it is just 30 **microns** in thickness, is translucent to the extent that its constituent minerals may be identified using their optical properties under a **petrographical microscope** (Fig. 1.30). Some minerals, though, do not let light through at all and are said to be **opaque**. These appear in thin sections as black areas, but may be studied instead by making **polished sections** and examining how they appear in reflected light (Fig. 1.31). Almost all of the metallic minerals, such as most sulphides and the native metals, are opaque, and **reflected light microscopy** is primarily used in studying the intergrowths and textures of metal-ores.

1.10 Mineral properties: specific gravity

Specific gravity is a numerical term that expresses the **density** of a mineral compared to that of water. It may be calculated by weighing a lump of a mineral and then lowering it into a beaker of water that is full to the brim, placed in a larger measuring beaker with a graduated side. The sample displaces a volume of water equivalent to its own volume, which may then be measured. Since a cubic centimetre of water has a mass of a gram, the volume of displaced water will be the same as its mass. To calculate the specific gravity, divide the mass of the mineral sample by the mass of the displaced water.

Let's run through a worked example. A lump of pyrite has a mass of 78 grams. Placed into the beaker, it displaces 15.6 cubic centimetres of water, in other words 15.6 grams. All you need to do is divide:

$$78/15.6 = 5$$

So pyrite has a specific gravity of 5. That's quite heavy, although there are much heavier minerals: gold has a specific gravity of 19.3. This represents one of the highest specific gravities in the mineral kingdom. Pure quartz has a specific gravity of just 2.65 – a cubic centimetre of it would have a mass of 2.65 grams. However, note that term, 'pure'. In reality, many minerals occur in nature as complex intergrowths involving several mineral species in small grains or crystals juxtaposed with one another. In such cases, measuring specific gravity may be extremely difficult, even in the laboratory. However, just by handling a lump of quartz or a gold nugget one can tell if it has a high specific gravity (very heavy for its size) or a low value (light for its size).

1.11 Mineral properties: hardness

Hardness, a product of the chemical bonding and structural properties of a mineral species, is an important characteristic that aids in identification and also determines the mineral's usefulness: for example, graphite is incredibly soft and has uses in lubrication, whilst corundum (Al_2O_3) is incredibly hard and has uses as an abrasive. The Mohs scale of mineral

hardness has been in use since the early nineteenth century and is a useful way of determining the hardness of a mineral in the field by determining whether it can scratch, or is scratched by, common household items like glass, copper coins, penknife-blades, porcelain and so on. Although it ranks minerals relative to one another, it is not an absolute scale – as the absolute hardness scale shows, topaz (8 on the Mohs Scale) is twice as hard as quartz but diamond (10 on the Mohs Scale) is sixteen times harder than quartz. Absolute hardness is an empirical value that has to be measured under fixed conditions: this is a laboratory task employing a piece of equipment known as a sclerometer, or a Vickers hardness-tester, which uses a diamond under a fixed load. When drawn across the surface of a mineral grain, the width of the resultant scratch is measured to determine this value. But for the field mineralogist, the Mohs scale is as useful today as it was almost 200 years ago. Each mineral on the scale will scratch those numerically below, and kits are available containing samples of the key minerals for use in the field.

1.12 Mineral species and their naming and classification

Earth – one of several rocky planets occurring in our Solar System – is, like its neighbours, almost entirely made of minerals, with only its Atmosphere and Oceans not counting as they consist of gases and liquid water. Every rock you pick up, every grain of sand on a beach, consists of one or more mineral species. At the time of writing, there are over 4500 known mineral species – each a distinct, individual element, alloy or compound. Every year, through new discoveries and research, the International Mineralogical Association (IMA) approves some 50–60 new mineral species to add to the list and there is a constant list of new species awaiting approval, once the requisite analytical work has been completed in order to fully characterize them. Once a mineral is approved as a species it is given a unique name and this is announced, complete with a full physical and chemical description, in a paper in the scientific literature.

The official IMA naming procedure has been in place since 1959, but, of course, some minerals have been familiar for very many centuries and retain their old, traditional, 'grandfathered' names. In both cases, a mineral's name may be based on where it was first found and described, such as cornwallite. It may reflect the mineral's composition: bismoclite, a bismuth oxychloride, has its name made up from parts of the names of its constituent elements. It may reflect some physical property: the name pyrite is based on the Greek *pyrites lithos*, meaning 'stone that strikes fire', which it most certainly does when struck with a hammer. It

Table 1.3 The Mohs/absolute scales of hardness

Mohs	Mineral	Absolute	In the field........
1	Talc	1	Easily scratched by a fingernail
2	Gypsum	3	Scratched by a fingernail
3	Calcite	9	Scratched by a copper coin
4	Fluorite	21	Easily scratched by a steel knife-blade
5	Apatite	48	Scratched by a knife-blade
6	Feldspar	72	Hardly scratched by a steel knife-blade but scratches glass
7	Quartz	100	Easily scratches glass or a steel knife-blade
8	Topaz	200	Scratches porcelain
9	Corundum	400	Easily scratches quartz
10	Diamond	1600	Will scratch everything else!

may reflect colour: azurite is named after its azure-blue colour, derived from an old Persian word. It may reflect the typical shape of a mineral's crystals – tetrahedrite usually crystallizes into tetrahedra. Or, and this is mostly a more recent development, it may celebrate a well-known mineralogist: thus russellite is a rare mineral named after the eminent English mineralogist Sir Arthur Russell (1878–1964).

The thousands of mineral species are divided up into a system of groups based upon their chemical composition. The classification is rooted in the fact that minerals are combinations of elemental or molecular ions: the classifications are based on the nature of the main atomic anion (such as sulphur, fluorine and so on) or complex molecular anion (such as phosphate or sulphate) – or, in the unique case of the native elements, the absence of an anion. There are a number of such classifications: for the sake of simplicity we'll look at the **Strunz-Nickel Grouping**, last revised in 2001, which divides the known minerals into ten basic groups as follows:

1 Elements (including metals and intermetallic alloys; metalloids and non-metals; carbides, silicides, nitrides, phosphides).
2 Sulphides and sulphosalts (including sulphides, selenides, tellurides; arsenides, antimonides, bismuthides; sulpharsenites, sulphantimonites, sulphbismuthites, etc.): the anions are atomic and comprise sulphur, selenium, tellurium, arsenic, antimony and bismuth – from groups 15 and 16 of the Periodic Table.
3 Halides (combinations of other elements with the atomic anions of the Group 17 elements fluorine, chlorine, bromine and iodine).
4 Oxides (also including the following combinations with molecular anions – hydroxides, V(5+6) vanadates, arsenites, antimonites, bismuthites, sulphites, selenites, tellurites, iodates).
5 Carbonates and nitrates (both combinations involving molecular anions).
6 Borates (combinations involving molecular anions).
7 Sulphates (also including the following combinations with molecular anions: selenates, tellurates, chromates, molybdates, wolframates).
8 Phosphates, arsenates, vanadates (all three being combinations involving molecular anions).
9 Silicates (combinations involving molecular anions).
10 Organic compounds (certain hydrocarbons, salts of organic acids etc.)

Each basic group is then divided into subgroups: for example, Group 1 is subdivided as follows:

1.A Metals and Intermetallic Alloys
1.B Metallic Carbides, Silicides, Nitrides and Phosphides
1.C Metalloids and Non-metals
1.D Non-metallic Carbides and Nitrides

A further subdivision then takes us to groups of similar minerals: Subgroup 1.A above then breaks down into:

1.AA Copper-cupalite family
1.AB Zinc-brass family
1.AC Indium-tin family
1.AD Mercury-amalgam family
1.AE Iron-chromium family
1.AF Platinum group elements
1.AG Platinum group element metal alloys
1.AH Miscellaneous elements, alloys

Finally, looking into subgroup 1.AA, we see the individual IMA-approved, named and fully described minerals that make up the subgroup, some of which are relatively widespread and may be seen in most mineral collections, others of which are extremely rare and perhaps limited in occurrence to microscopic grains from one locality only:

1.AA.05	Aluminium	Al
1.AA.05	Copper	Cu
1.AA.05	Electrum	(Au, Ag)
1.AA.05	Gold	Au
1.AA.05	Lead	Pb
1.AA.05	Nickel	Ni
1.AA.05	Silver	Ag
1.AA.10a	Auricupride	Cu_3Au
1.AA.10b	Tetra-auricupride	AuCu
1.AA.15	Anyuiite	$Au(Pb,Sb)_2$
1.AA.15	Khatyrkite	$(Cu,Zn)Al_2$
1.AA.15	Iodine	I_2
1.AA.15	Novodneprite	$AuPb_3$
1.AA.20	Cupalite	(Cu,Zn)Al
1.AA.25	Hunchunite	Au_2Pb

This is the case for each of the ten groups listed above, and today the classification is available online so that it can be explored at leisure. Entering 'strunz-nickel' into a search-engine will find several instances of the classification, and the author specifically recommends the Mindat website as a reliable source of information, where at the time of writing the classification may be explored starting at the following URL: http://www.mindat.org/strunz.php.

2 Typical mineral occurrences

2.1 Introduction

Typical mineral occurrences, often representative of the general abundances of elements in Earth's crust, may be classified along classic geological lines, representing the igneous, sedimentary and metamorphic rocks of the Earth's crust. To these may be added minerals brought up from the mantle by specific geological processes and the samples of other planetary bodies, including in some cases their inner zones, that are provided by the various classes of meteorites.

These occurrences of what are termed rock-forming minerals should be viewed in contrast with the atypical, highly to vastly enriched, concentrations of minerals in mineral deposits, whether economic (ore-grade) or sub-economic, which we shall examine in Chapter 3.

Some notes about the common rock-forming minerals are appropriate before discussion of each type of occurrence. Quartz (SiO_2) is almost ubiquitous in its occurrence in Earth's crust with the exceptions of ultrabasic igneous rocks and their metamorphosed equivalents. Other important rock-forming minerals occur as groups: that is, composition-ally-varying series with two or more pure end-members. In such groups, you have pure mineral 'X' at one end and pure mineral 'Y' at the other, with a series of minerals in between that contain X+Y in different proportions.

Such groups and their compositional determination are important in the classification of igneous and metamorphic rocks, so we will take a look at the six most important: the feldspars, micas, olivines, pyroxenes, amphiboles and garnets.

2.2 The feldspars

The feldspars are grey through white- to pinkish-coloured aluminosilicates with three pure end-members, namely potassium-feldspar (often annotated K-spar), with its two polymorphs, orthoclase (monoclinic) and microcline (triclinic), both $KAlSi_3O_8$; sodium-feldspar or albite, $NaAlSi_3O_8$ and calcium-feldspar or anorthite, $CaAl_2Si_2O_8$. There is a complete solid solution series between orthoclase and albite (i.e. feldspars with both potassium and sodium present in varying proportions): these are the alkali feldspars. Likewise, there is a complete solid solution series between albite and anorthite (i.e. feldspars with both sodium and calcium present in varying proportions): these are the plagioclase feldspars. There is no continuous solid solution series between K-spar and anorthite (Fig. 2.1).

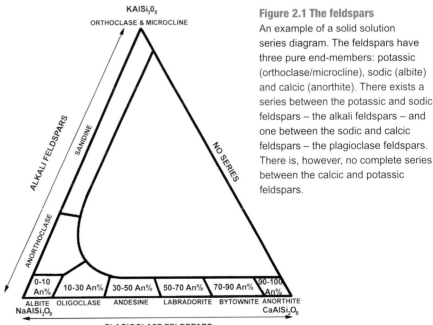

Figure 2.1 The feldspars
An example of a solid solution series diagram. The feldspars have three pure end-members: potassic (orthoclase/microcline), sodic (albite) and calcic (anorthite). There exists a series between the potassic and sodic feldspars – the alkali feldspars – and one between the sodic and calcic feldspars – the plagioclase feldspars. There is, however, no complete series between the calcic and potassic feldspars.

The alkali feldspar series is composed as follows:

Orthoclase (monoclinic): $KAlSi_3O_8$
Microcline (triclinic): $KAlSi_3O_8$
Sanidine (monoclinic): $(K,Na)AlSi_3O_8$
Anorthoclase (triclinic): $(Na,K)AlSi_3O_8$
Albite (triclinic): $NaAlSi_3O_8$

In thin section, the alkali feldspars often exhibit what is known as **perthitic texture**, which is an intimate lamellar intergrowth of sodic (e.g. albite) and potassic (e.g. microcline) feldspar resulting from **exsolution**, which is a process involving the unmixing of two minerals from one precursor mineral. Exsolution may often occur as a mineral assemblage cools and minerals stable only at high temperatures unmix into minerals that are stable at the lower temperatures now prevailing.

The plagioclase feldspar series (all triclinic) are classified in terms of their sodium–calcium ratio as 'Anx' where An is the proportion of anorthite composition present. Each species in the series is defined by a range of anorthite proportions. For example, the composition of oligoclase $((Na,Ca)[Al(Si,Al)Si_2O_8])$, is expressed as An10-30 – it is a sodium-rich, calcium-poor plagioclase. The whole plagioclase series is composed as follows:

Albite (An 0-10): $NaAlSi_3O_8$
Oligoclase (An10-30): $(Na,Ca)[Al(Si,Al)Si_2O_8]$
Andesine (An30-50): $(Na,Ca)[Al(Si,Al)Si_2O_8]$
Labradorite (An50-70): $(Ca,Na)[Al(Al,Si)Si_2O_8]$
Bytownite (An70-90): $(Ca,Na)[Al(Al,Si)Si_2O_8]$
Anorthite (An90-100): $CaAl_2Si_2O_8$

Exsolution may likewise occur in the intermediate plagioclase feldspars, but the resulting textures are, by contrast with the alkali feldspars, very fine-grained: however, refraction of light between such micro-lamellae gives rise to the bluish shimmer of labradorite that gives the mineral – or rocks containing it – many ornamental uses.

2.3 The micas
The mica group of sheet silicates are important rock-forming minerals that have the general formula:

$$X_1Y_{2-3}(Z_4O_{10})(OH,F)_2$$

in which X is potassium, sodium or calcium (less commonly barium, rubidium or caesium), Y is aluminium, magnesium or iron (less commonly manganese, chromium, titanium or lithium) and Z is silicon and aluminium (rarely ferric iron or titanium). Like the feldspars, the mica group contains a number of solid solution series: thus there is a partial series between the pale-coloured potassium and sodium end-members muscovite, $KAl_2(AlSi_3O_{10})(OH)_2$ and paragonite, $NaAl_2(AlSi_3O_{10})(OH)_2$ respectively and there is the biotite group of dark-coloured micas (biotite no longer being regarded as a discrete mineral species), which likewise extends between the magnesium and iron end-members phlogopite, $KMg_3(AlSi_3O_{10})(F,OH)_2$ and annite, $KFe_3(AlSi_3O_{10})(OH)_2$. In the field, most geologists still use muscovite and biotite as terms to describe any pale- or dark-coloured micas respectively: determining where on the series any pale or dark mica sits is not something that can be readily done outside of the laboratory. All micas crystallize in the monoclinic system and they have a perfect basal cleavage resulting from their sheet-like structure, so they appear in rock samples as lustrous flakes that readily catch the eye, especially when coarse-grained.

2.4 The olivines
The **olivine** group of orthorhombic rock-forming silicates, named after their greenish colour, consists of a solid solution series between magnesium (forsterite, Mg_2SiO_4) and iron (fayalite, Fe_2SiO_4) end-members. The compositions of olivine are expressed as molar percentages of these minerals, for example Fo70Fa30. The properties of olivine depend on the Mg–Fe ratios: for example, forsterite has a high melting temperature at atmospheric pressure (nearly 1900°C), but the melting temperature of fayalite is much lower (approximately 1200°C) and the melting temperature varies smoothly between these two end-members. Apart from iron and magnesium, some olivines may contain manganese or nickel, occurring in substitution for the standard metals.

2.5 The pyroxenes
The **pyroxenes** are a group of dark-coloured (greens and browns typically) rock-forming silicate minerals found in many igneous and metamorphic rocks. They share a common structure consisting of single chains of silica tetrahedra (for this reason they are termed **inosilicates**) and they crystallize in the monoclinic (clinopyroxene) and orthorhombic (orthopyroxene) systems. Pyroxenes have the following general formula:

$$XY[(Si,Al)_2O_6]$$

in which X is calcium, sodium, ferrous iron and magnesium (less commonly zinc, manganese or lithium) and Y is chromium, aluminium, ferric iron, magnesium and manganese (less commonly scandium, titanium or vanadium). The name pyroxene has an interesting source: it comes from two Greek words that combined give the term 'fire stranger'. The reason behind this was that pyroxene crystals were often found embedded in volcanic glass (rapidly cooled, 'quenched' lava) and were thus assumed to be impurities, hence the 'stranger' bit. In fact, the correct interpretation of this observation would most likely be that the pyroxenes had crystallized in the magma before it was erupted.

Common clinopyroxenes include augite, $(Ca,Na)(Mg,Fe,Al,Ti)[(Si,Al)_2O_6]$ and diopside, $CaMg[Si_2O_6]$. As with the other silicate groups, solid solution series occur, for example between the orthopyroxenes enstatite, $Mg_2[Si_2O_6]$ and ferrosilite $Fe_2[Si_2O_6]$ with most examples being of intermediate composition.

2.6 The amphiboles

The amphiboles (name derived from the Greek word for ambiguous!) are broadly similar to the pyroxenes in appearance (although some amphiboles have a strong tendency to form fibrous crystals) and, like the latter, have both monoclinic and orthorhombic members. The easy way to tell them apart in hand specimens is to look at their cleavage: amphiboles display cleavage planes intersecting obliquely, at about 120°, whilst pyroxenes have cleavage planes intersecting pretty much at right angles. Structurally, amphiboles are made up of double chains of silica tetrahedra; in terms of composition they differ from the pyroxenes in that they contain essential hydroxyl (OH) or halogen (F, Cl).

The general formula is as follows:
$$XY_2Z_5((Si,Al,Ti)_8O_{22})(OH,F,Cl,O)_2$$
in which X is sodium, potassium or calcium, Y is sodium, magnesium, ferrous iron and calcium (less commonly lithium or manganese) and Z is sodium, magnesium, ferrous or ferric iron, aluminium or chromium (less commonly lithium, manganese, zinc, cobalt, nickel, vanadium, titanium or zirconium).

Common amphiboles include actinolite, $Ca_2(Mg,Fe)_5Si_8O_{22}(OH)_2$, tremolite, $Ca_2(Mg_5)Si_8O_{22}(OH)_2$ and the hornblende solid solution series, which runs from ferrohornblende, $Ca_2(Fe_4Al)(AlSi_7O_{22})(OH)_2$ to magnesiohornblende, $Ca_2(Mg_4Al)(AlSi_7O_{22})(OH)_2$.

2.7 The garnets

Garnets are rock-forming silicates having the general formula $X_3Y_2(SiO_4)_3$, in which X is divalent calcium, magnesium or iron and Y is trivalent aluminium, iron or chromium in an octahedral/tetrahedral framework with $[SiO_4]^{4-}$ occupying the tetrahedra. They crystallize in the cubic system, and are most often found in the 12-faced dodecahedral crystal habit. They are hard minerals and have no cleavage. Because of the wide range of compositions, garnets display a wide range of colours, of which the most familiar are various shades of red, but orange, yellow, green and pink varieties are frequently met with. The cubic structure of garnets makes them relatively stable within high-pressure environments, which in turn means that they are frequently found in high-grade metamorphic rocks and in Earth's mantle.

Examples of garnets include almandine, $Fe_3Al_2(SiO_4)_3$, spessartine $Mn_3Al_2(SiO_4)_3$, andradite, $Ca_3Fe_2(SiO_4)_3$ and the spectacular bright green chromium-rich uvarovite, $Ca_3Cr_2(SiO_4)_3$ (Fig. 2.2).

2.8 Typical mineral occurrences
2.8.1 Minerals from space (meteorites)

Meteorites are rocks that are of extraterrestrial origin, having fallen to Earth's surface after being captured by the planet's gravitational pull. Exploring their origin takes us right back to the early years of the Solar System, when the planets were accreting together from the dust of the Solar Nebula. The accretion process formed both the rocky planets such as Earth and the gas-giants such as Jupiter, and in both cases the young planets captured the vast majority of the debris within their orbits early on. However, between Mars and Jupiter, the proto-planetary clumpings of matter were never able to coalesce into one planet due to the perturbation of Jupiter's gravitational pull, so that they remained as a scattered band of rocky debris from dust-size all the way up to the 950 km diameter Ceres, forming what is known as the Asteroid Belt. Some of the larger proto-planets underwent melting early in their existence, allowing the process of planetary differentiation to take place, with heavier elements sinking towards the centre of gravity and lighter elements rising towards the surface.

Gravitational perturbations also caused numerous collisions between individual asteroids and ejected much material from the Asteroid Belt during its early existence, some of which was flung in towards the inner planets, resulting in what is referred to as the Late Heavy Bombardment that

Figure 2.2A Almandine
Garnets vary in colour according to what transition metals they contain. Almandine contains iron and its crystals tend to be red or reddish-brown, as in this large hand-specimen with crystals up to 30 mm from Norland, Norway.

Figure 2.2B Uvarovite
Uvarovite contains chromium and is a spectacular bright green, as this detail from a hand-specimen covered in crystals to 3 mm, from the Ural Mountains in Russia, shows.

occurred around four billion years ago. This storm of meteors and asteroids cratered Earth heavily, although **plate tectonics** has since obliterated much of the evidence. On bodies where plate tectonics never became established, such as Mercury or the Moon, the cratering, much of it dating from this violent episode, is only too evident. Material is still ejected from the Asteroid Belt, and at the time of writing there are known to be 1330 asteroids more than 100 m in diameter whose orbits bring them close to that of Earth, although none are currently known to be on a collision course. Smaller bodies do frequently collide with Earth, such as the spectacular example in Russia in February 2013, and when they make it all the way to the ground without burning up or disintegrating they provide us with samples of the early Solar System, including the Asteroid Belt, the Moon and planets such as Mars: the very rare Martian meteorites must have mostly been blasted clear of the surface of that planet by massive impacts during those early and turbulent years.

Much material that enters Earth's atmosphere from space is so fine-grained that it burns up to produce the shooting-stars that are visible on many clear nights; the larger meteors that do reach the surface are notable events, producing bright fireballs that may be accompanied by sonic booms and loud explosions as they break up. Impact areas, when identified, are often descended upon by both academic and commercial interests searching for fragments across the **strewnfield**, the area in which the trajectory of the fireball suggests that specimens may be located. There is a considerable trade in meteorite samples, with rarer types commanding high prices.

Once collected, meteorites are classified by their compositions and textures: in some cases (such as iron meteorites) the classification may be fairly obvious, but cutting and polishing slices followed by microscopic examination and chemical analyses is key to their determination, especially in cases where meteorites from ancient falls have weathered outer zones. The basic classification is threefold, comprising stony, iron and stony-iron meteorites.

The **stony meteorites** consist mostly of rock-forming silicate minerals such as feldspar and olivine, with a range of accessory minerals. The commonest rocky meteorites are the **chondrites**, which make up about 85% of recorded falls. These contain small spheroidal mineral bodies – the chondrules – which include metallic nickel and iron, sulphides and other minerals. An important if uncommon subgroup, the carbonaceous chondrites, also contain water and a range of organic compounds; the presence of these volatiles indicating that they have not undergone significant heating since accretion, this leading to the hypothesis that their composition may be fairly close to that of the Solar Nebula from which the planets originally accreted.

The **achondrites** are relatively uncommon (*c*.8% of recorded falls) stony meteorites that typically consist of basic igneous rock (such as basalt or gabbro); these include the meteorites that are thought to have originated from the surface of the Moon or Mars.

The **iron meteorites** are made up primarily of the crystalline iron-nickel alloys kamacite and taenite, together with accessories such as the iron sulphide troilite, the Fe-Ni phosphide schreibersite and the Fe-Ni-Co carbide cohenite. They only comprise some 5% of recorded falls but are disproportionately represented in many collections, being much easier to recognize by lay people, more resistant to weathering, and more likely to make it down to Earth's surface. The irons are divided into three subtypes: firstly the hexahedrites (<6% Ni), secondly the octahedrites (6%–17% Ni) and thirdly the ataxites (>17% Ni). Of these perhaps the best known, because they are frequently displayed in collections, are the octahedrites, consisting of a regular crystalline lamellar intergrowth, known as the **Widmanstätten pattern**, of kamacite and taenite (Fig. 2.3). This distinctive texture is revealed by etching the surfaces of polished slices of octahedrites (taenite being more resistant to acids than kamacite). The formation of such coarsely crystalline metal intergrowths can only take place during very slow cooling over millions of years, and is thought to have occurred in the cores of larger asteroids/planetoids that had undergone melting and differentiation, before being shattered by major collisions with other large bodies. It is likely that the metallic core of Earth, which we cannot directly sample, of course, although we can calculate its density and likely composition, is compositionally similar to the iron meteorites.

The stony iron meteorites consist of a mixture of kamacite, taenite, troilite, cohenite, schreibersite and non-metallic minerals. Relatively rare, they constitute just over 1% of meteorite falls.

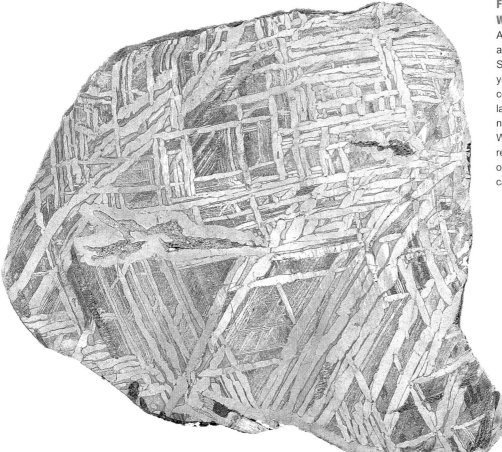

Figure 2.3 Iron meteorite with Widmanstätten pattern
A slab, 15 cm wide, cut from an iron meteorite found at Seymchan, Russia, some years ago. It displays a texture consisting of a regular crystalline lamellar intergrowth of different nickel-iron alloys, known as the Widmanstätten pattern, which is revealed by polishing the surface of the slice and then etching it carefully in acid.

They are divided into two subtypes, the **pallasites**, consisting of large olivine crystals set in a nickel-iron matrix, and the **mesosiderites**, which consist of intergrown silicates and metal grains. Pallasites are a familiar sight in many museums, displayed backlit so that the gem-quality olivine crystals stand out (Fig. 2.4). There are two lines of thought as to their origin: one is that they represent fragments of the core–mantle boundary of a differentiated large asteroid/planetoid; the other is that they

are impact-generated mixtures of core and mantle materials. Mesosiderites are even more uncommon and their breccia-like internal textures have led to the hypothesis that they have been formed by accretion following violent large asteroid/planetoid collisions.

2.8.2 Minerals from within the Earth's mantle

As is the case with Earth's **core**, we cannot directly sample the depths of its mantle although we can investigate

its physical properties via geophysical methods. Through geophysics – specifically by the way that seismic waves behave as they pass through it – the mantle has been divided into three zones: the **upper mantle**, starting at the base of the crust and extending down to 410 km, the transition zone, from 410 down to 660 km and the **lower mantle**, which continues on down to the core–mantle boundary at 2890 km. Through deep-seated geological processes, we can, albeit very locally, obtain samples

Figure 2.4 Pallasite
Another prepared specimen from Seymchan, Russia, in which large olivine crystals are intergrown with the nickel-iron alloys. Such meteorites are termed pallasites.

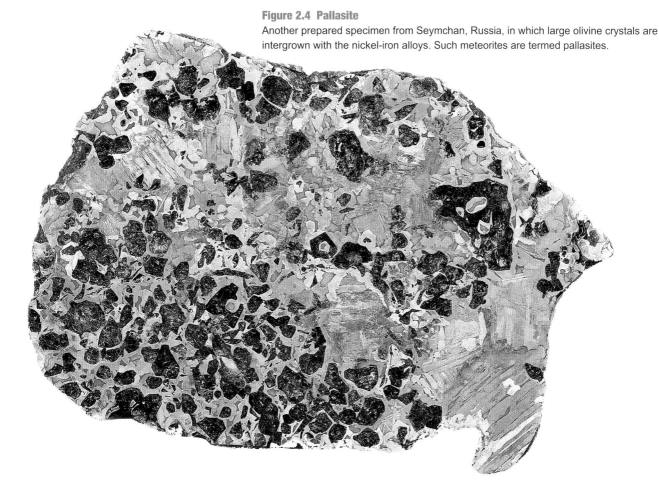

from some of these zones, brought up rapidly from great depths. These occur in the form of **xenoliths** and **xenocrysts** – exotic rock fragments and crystals respectively, captured and transported by magmas and then entrapped within them upon solidification into intrusive rocks such as lamprophyres and **kimberlites**. The latter are famous as the key primary source of diamonds. Along with these precious stones, many other mantle-derived minerals are high-pressure polymorphs: minerals

with identical chemical compositions to other species that form in the crust but with a different (typically as compact as possible) lattice configuration that remains stable under the immense pressures prevailing at great depths.

As an example, magnesium iron silicate $((Mg, Fe)_2SiO_4)$ is an abundant component of the mantle. In both the upper part of the mantle and the crust, it occurs in the form of olivine, which crystallizes in the orthorhombic system. Two high-pressure polymorphs

of olivine are known: wadsleyite (orthorhombic) and ringwoodite (cubic) system: in nature, these have, until recently, only been known from meteorites. However, both minerals have long been inferred to occur, with increased depth, in large quantities in the transition zone, with wadsleyite the stable phase in its upper part and ringwoodite in its lower part. In 2014, this inferred occurrence was confirmed by researchers examining a Brazilian diamond: they identified microscopic inclusions of ringwoodite. Moreover, the research showed that the ringwoodite contained small quantities of water, something that had already been predicted experimentally. This additional discovery, taken together with the theoretical abundance of ringwoodite in the transition zone, has interesting connotations for the debate regarding the origin of water on Earth.

Diamonds are thought to crystallize at depths of at least 150 km down – and in the case of the ringwoodite-bearing Brazilian example, a lot deeper than that. It is difficult to imagine the pressures involved: in the upper mantle part of the diamond stability field, pressures are upwards from 6,000,000,000 Pa. For comparative purposes, standard atmospheric pressure at sea level is 101,325 Pa and at the bottom of the mantle the pressure is 140,000,000,000 Pa. Under the pressures that prevail within the diamond stability field, carbon crystallizes in the cubic system, mostly forming octahedra, contrasting sharply with the common near-surface forming carbon polymorph, graphite, which has a much less compact, layered hexagonal structure and is one of the softest of all minerals.

Capture of diamonds by kimberlite and lamproite magmas deep in the mantle is only the first part of a perilous journey. The movement of the magma up through the crust must be rapid, so that the diamonds are transported quickly to the near-surface site of magma solidification and then trapped as the rock cools. As with many other high-pressure minerals, if they are removed from their stability zone, for example from high temperature/ high pressure to high temperature/

low pressure, they may destabilize and revert to polymorphs that are stable at lower pressures. As a consequence of that, not all kimberlite pipes are diamond-bearing: the pipes tend to occur in clusters in which only a small proportion may be diamondiferous.

A frequent initial step in locating likely areas to prospect for lamproites or kimberlites – and thereby diamonds – is to examine a geologically-favourable area for unusual indicator minerals formed in the same high-pressure mantle environment. These will be the minerals making up the bulk of the lamproite or kimberlite: erosion and weathering of any outcrop of these rocks will inevitably lead to them occurring in soils or stream sediments. Potentially useful indicator minerals include chromium-bearing garnets such as bright red pyrope or green ugrandite, chromite (a mineral of the spinel group with the formula $FeCr_2O_4$) and the bright green pyroxene, chrome-diopside. The finding of several of these high-pressure minerals occurring together in, for example, panned heavy mineral concentrates from the stream sediments of a river catchment, would identify that area as prospective, and more elaborate and expensive exploration techniques, such as airborne geophysics, may then be justified. If, finally, a system of kimberlite pipes is discovered and found to be diamond-bearing, the grade still needs to be assessed: typically, kimberlite needs to contain 10 **carats** (2 grams) of diamond per hundred tonnes to be worth working – but of course the quality and size of any diamonds that are present is another important factor. An exceptional find can see

grades of hundreds of carats of diamonds per hundred tonnes, but even so, 500 carat ore still only means that a hundred tonnes will contain 100 grams of diamonds – thus emphasizing the scarcity of this mineral outside of its formational environment, hundreds of kilometres beneath our feet.

2.8.3 Minerals from igneous rocks

Igneous rocks are defined as those formed by the cooling and solidification of magma, molten rock produced at high temperatures beneath the Earth's surface. Their mineralogy is primarily a function of chemical composition, which is in turn a function of the magma source, for example **subduction**, high-grade metamorphism (**partial melting**), **mantle heat-plumes** and so on. Igneous rocks may be **extrusive** (erupted under water or subaerially) or **intrusive** (not erupted but emplaced within Earth's crust).

The textures of igneous rocks depend upon how quickly they are cooled: a magma intruded deep into the Earth's crust will cool more slowly (tens of thousands to millions of years) allowing crystals more time to grow and resulting in a coarser grain-size, visible to the naked eye (**phaneritic** rocks). In extreme cases, such crystals may be huge, resulting in the rock having a very coarse texture. In contrast, a magma erupted from a volcano will tend to cool very quickly, as a consequence of which the constituent minerals may be too fine-grained to observe without a microscope (**aphanitic** rocks). In extreme cases, magma may be cooled instantaneously, for example when erupted onto the sea-bed, the resultant **quenching** of the heat being so rapid that a natural volcanic glass forms at the interface between the magma and its coolant; there has simply not been enough time for the minerals to crystallize. Sometimes, complications arise in this simple scheme of things: a very common one is where a magma begins to crystallize, so that crystals of certain minerals form over tens to hundreds of thousands of years, but it is still liquid in bulk and able to flow to its final resting place as an intrusion, or to where it is erupted at surface. In such cases, well-developed and relatively large **phenocrysts** of the minerals that had originally started to crystallize are found in a much finer groundmass of the other constituent minerals: this is known as a **porphyritic** texture.

Igneous rocks are classified into four key groups, defined by their silica content. **Acidic rocks** contain more than 65% SiO_2 and are typically rich in 'light' elements such as sodium, potassium and aluminium. They tend to be light-coloured, their mineralogy dominated by quartz, alkali feldspars and micas. Slowly-cooled, coarsely crystalline intrusive acidic rocks are typified by the granites and granodiorites; their fine-grained eruptive equivalents are the rhyolites and dacites (Fig. 2.5).

Associated with many granites are granitic pegmatites (intermediate and

Figure 2.5 Examples of acidic igneous rocks
Acidic igneous rocks tend to have an overall pale colour, as in these two sections of drill-core from a granite in Cornwall, England.

basic pegmatites do occur but are less common). **Pegmatites** are late-stage features, forming pods within the granite and/or dykes and sills in the country-rock. Towards the end of the solidification process of an intrusion, the remaining fluid is enriched in water and a range of other volatile and incompatible elements that have not been incorporated into those minerals

Table 2.1 Igneous rocks: a simple classification

COMPOSITION SETTING/TEXTURE	Acidic	Intermediate	Basic	Ultrabasic
Intrusive: Phaneritic (including porphyritic)	Granite, Granodiorite	Diorite	Gabbro, Dolerite	Peridotite, Pyroxenite
Extrusive:Porphyritic & Aphanitic	Rhyolite, Dacite	Andesite	Basalt	Komatiite
Extrusive: Glassy	Obsidian, Pumice		Basaltic Glass	
Pyroclastic	Volcanic Tuff			

making up the bulk of the granite. The conditions for pegmatite formation appear to require a relatively low nucleation-rate and relatively rapid growth of any crystals that start to form, resulting in a low number of very large crystals: individual crystals several metres in their longest dimension have been recorded. Granite pegmatites primarily consist of quartz, feldspar and mica, but with a wide and highly variable accessory mineral suite that typically includes volatile-rich minerals such as fluorite, apatite and tourmaline.

Intermediate rocks are less silica-rich, containing 52–65% SiO_2 with less than 10% quartz: often, both alkali and plagioclase feldspars are present. Darker in colour than acidic rocks, they are often a pinkish-grey hue. Diorite is a widespread intermediate intrusive rock, whereas the erupted equivalent is andesite. **Basic rocks** are relatively silica-poor at 45–52% SiO_2: they are, however, relatively rich in calcium, iron, and magnesium, from which characteristic the alternative name, mafic rocks, is derived. The increased content of these heavier elements and the minerals that they contain such as amphiboles, pyroxenes and olivine gives basic rocks a dark, often grey to black colour; gabbros and dolerites are common intrusive basic rocks, whilst basalts are the eruptive equivalent (Fig. 2.6). Quartz may be present but in very minor quantities, whilst the feldspar content is entirely plagioclase. Within this class there are also the anorthosites – rocks made up almost entirely of plagioclase feldspar.

Basic rocks make up **oceanic crust**, the upper parts of which are erupted as **sub-alkaline** (relatively low alkali metals content) basalts known as

Figure 2.6 Basic igneous
Basic igneous rocks tend to be dark in colour, as in these two polished pieces of dolerite from North Wales

tholeiites at mid-ocean ridges. This composition differs from the oxidized alkali basalts commonly erupted at volcanic island-arcs, especially those sited upon **continental crust**. It reflects differences in magma formation: tholeiites are formed by partial melting of mantle material whereas in the case of alkali basalts, subduction and partial melting of subducted crustal slabs is also a factor.

Ultrabasic (or ultramafic) rocks contain less than 45% SiO_2 and they consist essentially of ferromagnesian minerals such as pyroxene, olivine and spinel (chromite and/or magnetite), with up to 10% of plagioclase feldspar. Peridotites are intrusive ultrabasic rocks consisting in some cases almost entirely of olivine: pyroxene may or may not be present. Komatiites are extrusive ultrabasic rocks confined in their occurrence to Archaean and (less so) Proterozoic strata that consist of olivine, pyroxene and small amounts of plagioclase and chromite.

Intrusive ultrabasic rocks often occur in large, layered bodies known as **cumulates**, where differentiated rock types occur in layers. Individual layers do not represent the bulk chemistry of the magma from which they crystallized: instead, certain minerals crystallized early during the cooling process (for example, because they had high melting-points) and sank down through the magma to form layers parallel to the floor of the chamber. A classic example is the Skaergaard intrusion in Greenland, where a 2500 m thick layered cumulate, thought to have crystallized from a single confined magma chamber, shows a distinct mineralogical and geochemical layering. Plagioclase varies from calcium-dominant near the base to sodium-dominant near the top; CaO averages 10.5% of the rock at the base, decreasing to 5.1% at the top, whereas $Na_2O + K_2O$ average 2.3% of the rock at the base, increasing to 5.9% at the top. Olivine varies from magnesium-dominant $(Mg,Fe)_2SiO_4$ near the base to the pure iron end-member fayalite (Fe_2SiO_4) at the top; MgO averages 11.6% of the rock at the base, decreasing to 1.7% at the top, whereas FeO averages 9.3% of the rock at the base, increasing to 22.7% at the top.

Then there are the oddballs, such as kimberlite (discussed above) and the **carbonatites**. The latter are unusual rocks made up of at least 50% carbonate minerals, that may occur as intrusions or lavas erupted at surface.

Carbonatite lavas are relatively cool, typically 500–600°C when erupted, compared, for example, to basaltic lavas at 1100°C or more. They are rare rocks and there is only one volcano erupting carbonatite lavas at the present day, in the East African Rift Valley.

In all the above classes of igneous rocks, a wide range of minor **accessory minerals** may be present: these include oxides, sulphides, halides, phosphates, carbonates and silicates. Many relatively rare metals and other elements may be represented, sometimes (albeit rarely) to the extent that the rocks can be considered to be a source of such commodities. These occurrences, however, are atypical: most igneous rocks are merely sources of ornamental or building stone or aggregate, so that igneous-related metalliferous mineral deposits will be discussed in the next chapter.

2.8.4 Minerals from sedimentary rocks

Sedimentary rocks are those formed on the surface of the Earth's crust via physical and chemical processes of **erosion** and weathering, transport as detritus and ions in solution, deposition and precipitation and finally **diagenesis**, by which a sediment becomes converted (lithified) into a rock. They may form on land or under water: all that is required is a source for the sediment (or ions) and a place of deposition or precipitation.

Because of the above parameters, sedimentary rocks vary widely in their detailed mineralogy, but nevertheless the rocks themselves have a straight-forward basic classification which we shall now examine. The main divisions are as follows: firstly, there are the **clastic** sedimentary rocks– those

consisting primarily of detritus; secondly, chemical precipitates – rocks formed by the precipitation from solution of minerals, for example **limestones**; thirdly, **evaporites** – rocks formed when minerals crystallize due to bodies of water evaporating away; and fourthly, organic deposits such as coal, derived from decomposed plant life.

Clastic sedimentary rocks are made up, as the name suggests, from clasts – grains of detritus varying in size from a few microns to hundreds of metres across, depending upon the environment of deposition and the physical processes operating at the time. The detritus originates via the erosion, by the actions of wind, water and ice, assisted by gravity, of pre-existing rocks of all types. There is a cyclic nature to this story: an igneous rock may be eroded to generate sediment which then accumulates, solidifies into a sedimentary rock and may later undergo metamorphosis into a metamorphic rock, which may later be remelted into a magma, or in time be eroded once again to form a new sedimentary rock. It all depends on the geological history of a given area.

The clastic sedimentary rocks include (from coarser to finer) **conglomerates**, **sandstones**, **siltstones** and **mudstones**. The bulk of minerals making up clastic sedimentary rocks are common and relatively weathering-resistant rock-forming species, such as quartz, feldspar, mica and clay-minerals. However, both common and minor minerals in an eroding rock may end up in the resulting sediment, a useful thing as some chemically-stable and hard minerals, such as zircon, can be used in research into the **provenance** of ancient sediments – in other words, they can indicate where the original, eroding source-rocks were. This can be of great use in palaeogeographical reconstructions. All such minerals, eroded from source and then transported as detrital grains to a depositional environment, are referred to as **allogenic**, in contrast to **authigenic** minerals which have precipitated from solution *in situ* (Fig. 2.7).

Working in combination, gravity and water (or sometimes wind) action can be highly efficient at sorting mineral grains according to their density, something that Man has known about for a very long time, and a physical principle that was behind the invention of the gold-pan. Where ideal conditions prevail, denser minerals may become concentrated as they sink down through the lighter sediment, which is progressively winnowed away by the current. Once they reach an impenetrable layer, such as the bedrock floor of a stream, they may begin to accumulate to form placer deposits, potentially worth exploiting, which are discussed in the following chapter.

Chemical precipitates are sediments that have formed by the precipitation of dissolved substances from water. Of these, the most common worldwide are limestones, deposited by the precipitation of calcium carbonate (calcite). Magnesium can sometimes substitute for the calcium in calcite to form the mineral dolomite, $CaMg(CO_3)_2$. In such cases, a limestone may be described as dolomitic. In rather stagnant sedimentary environments, with abundant organic matter present, iron carbonate (siderite, $FeCO_3$) may accumulate, producing clay-ironstone beds.

Limestone is a major global

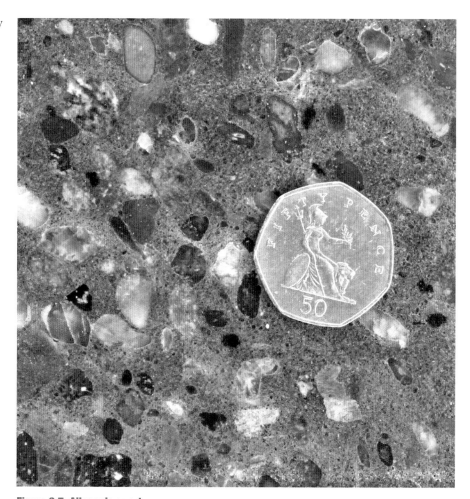

Figure 2.7 Allogenic quartz
The quartz pebbles in this Cambrian conglomerate from North Wales are allogenic; i.e. they are mineral grains that have been eroded from source and then transported as detrital grains to a depositional environment.

carbon-sink, and as such plays a vital role in the climate-controlling carbon cycle. Carbon dioxide, dissolved in rainwater to form a weakly acidic solution, is an important weathering agent. The solution is able to attack certain silicates – and carbonates – dissolving them gradually and transporting their constituents as ions through drainage systems into lakes or the sea. Here, chemical reactions allow the calcite to precipitate out: the precipitation process may occur directly from water or involve various organisms such as the corals, molluscs and forams that secrete calcium carbonate in order to make their shells (Fig. 2.8). Thus, some limestones may be made entirely from fossils. Either way, enhanced phases of weathering, perhaps following a major

Figure 2.8 Fossilised corals
Marine organisms such as corals make their skeletons by precipitating calcium carbonate out from seawater. These colonial corals are of Lower Carboniferous age and are from North Wales.

episode of volcanism or mountain-building, represent major drawdowns of atmospheric carbon dioxide, and there is some evidence for such phases of activity in the past having led to significant global cooling episodes.

Evaporites, which form when saline solutions, for example in salt-lakes or enclosed seas, are evaporated away, may entirely consist of water-soluble minerals, the most familiar being rock-salt (or halite, NaCl) and sylvite (KCl). Other minerals such as gypsum (CaSO$_4$·2H$_2$O) are well known from evaporitic rocks, but in fact there are many dozens of mineral species known from such rocks. Non-marine evaporites may have differences in their miner-alogy from those formed in marine environments, because of differences in the original water chemistry. In the UK, evaporite deposits occur within red-bed sequences formed in hot, arid conditions during Permo-Triassic times: they have been mined historically and

Figure 2.9 Evaporites
Red Triassic mudstones on the South Wales coast were deposited in arid conditions where lakes periodically evaporated. The white band in the lower part of the cliff is an evaporite horizon, formed by just such a process, and it consists mainly of gypsum.

continue to be exploited today (Fig. 2.9).

Organic sedimentary rocks – coal and other related deposits – are formed *in situ* from decaying vegetable matter in stagnant, reducing conditions. Such conditions also favour the deposition of metal sulphides, and many coals are enriched in a variety of metals, although by far the commonest accessory mineral occurring in coal is pyrite, iron sulphide.

Diagenesis is the term that covers all the processes (physical, chemical and in some cases microbial) undergone by any sediment following its deposition, subsequent progressive burial and transition into solid rock. Increasing pressure consequential upon increasingly deep burial forces increasingly more trapped **pore-water** from the sediment. Minerals dissolve and recrystallize, or are replaced by other minerals, to form **cements** that bind the sediment's constituent grains. The chemistry of such processes determines how hard, or how resistant to weathering, a rock will be: for example, a sandstone cemented by silica will be both very tough and weathering-resistant, whereas one in which the cement includes carbonates will be softer and more prone to weathering. Temperature is also of importance during diagenesis, and at steadily increasing pressures and temperatures diagenesis overlaps into **metamorphism**: in a sense, diagenesis is the lowest grade of metamorphism possible.

As mentioned above with respect to coal, some sediments contain anomalous concentrations of various metals. During diagenesis, the metals may combine with sulphur to form sulphide minerals, common examples being pyrite cubes that are commonly found in mudstones or **pyritized** fossils such as ammonites. Sometimes, sulphides, carbonates, phosphates or silica can grow into **concretions**, rounded bodies of mineralized rock that may, due to their relative hardness or resistance to

weathering, stand out from outcrop surfaces or remain behind when a mudstone outcrop has otherwise been eroded away, for example by wave-action. Concretions can sometimes have an internal structure consisting of a network of cracks dividing them up internally into sections: such concretions are usually referred to as **septarian nodules**. The cracks (or septa) may be completely filled or contain open cavities lined by a range of well-crystallized minerals. A famous example in the UK is in the South Wales Coalfield, where the cavities in the clay-ironstone concretions contain excellent examples of crystallized millerite, a nickel sulphide (NiS) (Fig. 2.10). The attendant well-formed quartz crystals contain fluid inclusions that have yielded

important data regarding the pressures and temperatures of concretion development (and thereby the conditions of diagenesis in the coalfield).

2.8.5 Minerals from metamorphic rocks

The term '**metamorphic**' comes from the Greek, and it generically means 'to change form' – such as the metamorphosis from caterpillar to pupa to adult butterfly. In geology, it specifically refers to the changes that take place in the solid state due to variations in pressure and temperature conditions, which cause original sedimentary or igneous rocks to undergo mineralogical changes. These may involve recrystallization of the existing minerals, resulting in textural changes. Prominent

Figure 2.10 Millerite
Golden metallic needles of millerite, with quartz and clay-minerals, in a siderite-lined cavity in a clay ironstone nodule from the South Wales Coalfield. This distinctive mineralization is thought to have formed during the diagenesis of the coal-bearing strata.

examples include the pressure-aligned recrystallization of sheet-silicates like clay-minerals and micas, leading to the development of schistosity and, in mudstones, a slaty cleavage. More intense metamorphism can lead to a partial to complete change in the mineralogy. Very low-grade metamorphism is essentially just an extension of diagenesis, whereas the other end of the scale of intensity is reached at very high temperature conditions, beyond which the rock melts into a magma.

Metamorphism can affect large areas of crust as an effect of continental collision, mountain-building, subduction and so on (**regional metamorphism**) or more localized areas such as the contact-zones of igneous intrusions (**contact metamorphism**), shear-zones and other fractures (**dynamic metamorphism**) and impact craters (**shock metamorphism**).

Regionally-metamorphosed rocks are indexed into **facies** (these being rocks with specific sets of characteristics) by the **mineral assemblages** they contain. Metamorphic facies reflect pressure (P) and temperature (T) conditions during the metamorphism: these are referred to as low (L), medium (M) and high (H). Starting with the lowest grade, zeolite facies (LP/LT) metamorphism leads to the development of hydrated silicate minerals belonging to the zeolite group, such as laumontite, $Ca(AlSi_2O_6)_2 \cdot 4H_2O$, associated with albite, mica, chlorite and clay minerals.

The next step up is the prehnite-pumpellyite-facies (LP/LT), a similar overall assemblage, but in which the place of the zeolites is taken by prehnite, $Ca_2Al(AlSi_3O_{10})(OH)_2$ and minerals of the pumpelleyite group, such

as $Ca_2MgAl_2(SiO_4)(Si_2O_7)(OH)_2 \bullet (H_2O)$. Epidote, a distinctive pistachio-green mineral with the formula $Ca_2Al_2(Fe^{3+},Al)(SiO_4)(Si_2O_7)O(OH)$, is often present as part of this facies, especially where basic igneous rocks were the unmetamorphosed precursor.

Moving up into the middle and higher grades, we firstly have greenschist facies (MP/MT): as well as quartz, mica and albite these rocks contain epidote and the amphibole actinolite,

$Ca_2(Mg,Fe)_5Si_8O_{22}(OH)_2$; in fine-grained **metasediments** (collectively referred to as pelites) the manganese-bearing garnet, spessartine, may be present. Amphibolite facies (MP/MT-HT) rocks typically contain garnets and amphiboles such as members of the hornblende group plus plagioclase, quartz and mica; less commonly blue kyanite, Al_2SiO_5, and red-brown staurolite, $Fe^{2+}_2Al_9O_6(SiO_4)_4(O,OH)_2$ may be present (see Fig. 2.11). Granulite facies (MP/

Figure 2.11 Kyanite and staurolite in mica-schist
High-grade (amphibolite facies) metamorphic rock from the Central Alps of Switzerland containing long prisms of blue kyanite and red-brown staurolite set in a micaceous matrix. The rock was originally a mudstone.

HT) metamorphism, the highest grade of regional metamorphism away from subduction zones, leads to assemblages containing garnets, ortho- and clino-pyroxene, amphiboles plus cordierite, $(Mg,Fe)_2Al_4Si_5O_{18}$, sillimanite (a polymorph, with kyanite, of Al_2SiO_5) or sapphirine, $(Mg,Al)_8(Al,Si)_6O_{20}$ (depending on the type of precursor rock).

In subduction zones, where cold slabs of wet oceanic crust are forced down many kilometres into the mantle, it is pressure that is the major driver of metamorphism at first, forming blueschist facies (M-HP/LT) rocks from $c.15$ km down and then, at greater depths (from $c.80$–120 km down), the high-density eclogite facies (HP/HT). Blueschists are so named because of the presence of the bluish-grey amphibole glaucophane, which is often accompanied by lawsonite, $CaAl_2Si_2O_7(OH)_2 \cdot H_2O$, plus chlorite and a range of other minerals. Eclogites are distinctive rocks containing abundant bright green omphacite – a pyroxene – and red garnets, plus various accessories that can include diamond (Fig. 2.12). They are relatively uncommon at the Earth's surface, because to get back to surface unaltered they need to be rapidly exhumed, a process that requires many tectonic boxes to be ticked! Often, the process is so slow that **retrograde metamorphic** reactions occur, i.e. they recrystallize from a HP/HT facies to a MP/MT facies such as amphibolite.

Contact metamorphism essentially

Figure 2.12 Eclogite
Eclogite from south-western Norway. This was once oceanic crust! Subducted into the mantle and later exhumed through tectonic activity, it comprises an assemblage of strikingly colourful minerals, dominated by the bright green pyroxene, omphacite, and abundant red garnet.

occurs at low pressures and high temperatures, often quite high up in the crust. The overall result is known as **hornfels**, a typically grainless, splintery rock, but which may along joints contain well-crystallized examples of its constituent minerals. In addition, euhedral crystals of certain minerals may grow *in situ* within the host rock – these are known as **porphyroblasts**. The contact-metamorphosed zone surrounding the intrusion is termed the **metamorphic aureole**, in which the contact-metamorphic effects are greatest adjacent to the intrusion, decreasing in intensity away from it. The size of the aureole is dependent on a number of variables: the size of the intrusion and its temperature are clearly major factors, but the temperature difference between it and the host rocks is important too.

Just as with regional metamorphism, the resultant mineral assemblages change with increased heat. At low to medium temperatures, they contain quartz plus micas, albite, epidote and actinolite; with increasing temperatures amphiboles, plagioclase, cordierite and andalusite, (another polymorph, with kyanite and sillimanite, of Al_2SiO_5). At medium to high temperatures, orthopyroxenes and sillimanite appear, as does garnet, whilst at very high temperatures the high-temperature alkali feldspar, sanidine, is found, often accompanied by glass formed by partial melting.

A particularly important form of contact-metamorphism occurs where magma intrudes limestone sequences and other carbonate-bearing rocks: the result is a rock containing, as well as recrystallized carbonates, the olivine forsterite, the pyroxene diopside, the garnet grossularite and many other mineral species. Such rocks are often attractive when unweathered, are compact, and so take a good polish and are quarried for their use as an ornamental stone – in other words, marble.

Dynamic metamorphism sums up a range of processes that occur in zones of generally high tectonic strain, such as major **basement**-reaching faults, thrust- and shear-zones. Rocks subjected to dynamic metamorphism alter their appearance due to textural changes: at shallow depths of up to a few kilometres a zone of **cataclasis** is formed, where the rock is simply shattered and sometimes ground into powder. At greater depths, at greater temperatures and pressures, the rock becomes relatively ductile and deforms by shearing in bulk, forming a streaky rock known as a **mylonite**.

Finally, shock metamorphism is a relatively infrequent process that occurs when a sizeable meteorite or asteroid collides with the Earth's surface, a process typically involving sudden heating to high temperatures and massive pressures. Certain minerals are diagnostic of such environments, such as the silicon dioxide polymorphs coesite (monoclinic) and stishovite (tetragonal) and microscopic diamonds.

3 Atypical concentrations of minerals

3.1 Introduction

While the typical mineral occurrences described in Chapter 2 often broadly reflect the crustal abundances of the elements that constitute the individual minerals, all minerals also occur in more localized but sometimes very high concentrations. These are termed mineral deposits, and if sufficiently rich to extract at a profit, the term ore-deposit may be applied.

Mechanisms for concentrating minerals are as varied as the deposits that they form, and include igneous, sedimentary and metamorphic processes, the effects of weather and climate and even biochemical reactions. In essence, all that is required is a site where there is room for a concentration to form, a means of sourcing and transporting elements to it, and then a means by which the minerals may be deposited. Sounds simple? It is, in the most basic terms, but as this chapter will reveal, it covers a remarkable diversity of deposits and their formational processes.

3.2 Igneous environments

3.2.1 Magmatic cumulate mineral deposits

Some internationally important ore-deposits owe their existence to processes that took place deep underground within magma-chambers. Over time, within this molten environment, in a process known as magmatic segregation, metalliferous and other minerals form, either as crystals or immiscible liquid drops (think oil and water as an analogy for the latter) and, according to gravity and their density relative to the parent magma, either settle downwards or float upwards, to become accumulated in a layered fashion. Metalliferous cumulates occur almost exclusively in mafic or ultramafic intrusions, which in general tend to be enriched in certain metals like nickel or chromium.

In terms of tectonic setting, the majority of the world's biggest cumulate mineral deposits occur at sites of intra-crustal rifting, although a fascinating exception is also one of the world's most important nickel-producers, at Sudbury in Ontario, Canada. Here, metals are obtained at a number of mines, all of which exploit ore-bodies formed in the lower layers of a segregated basic crustal melt-sheet, formed in the immediate aftermath of a significant (~10 km wide object) extra-terrestrial impact, nearly two billion years ago, that left an elliptical impact structure some 60 x 30 km in size.

The segregation process varies according to the prevailing physical conditions within the magma chamber over time: three types of cumulates are recognized as a result. Adcumulates may consist entirely of accumulated minerals; mesocumulates have more than 85% accumulated minerals and orthocumulates have more than 75% accumulated minerals occurring in a groundmass formed by crystallization of the remaining magma. Prefixed to these terms are the names of the most prominent minerals present: thus a layer containing 82% olivine, 7% magnetite and the remaining 11% the groundmass would be referred to as an olivine-mesocumulate.

Light minerals that float up to form cumulate layers include feldspars, which in some layered intrusions occur as the pale rock anorthosite (defined as a rock containing more than 90% plagioclase feldspar). In some cases, anorthosite layers may be quarried as a source of plagioclase for various industrial purposes. Heavy minerals include dense silicates like olivines, oxides such as magnetite (Fe_3O_4), ilmenite ($FeTiO_3$), chromite ($FeCr_2O_4$) and sulphides. Some of the resultant cumulate rocks contain high concentrations of metals: chromitite, for example, is a rock containing more than 90% chromite. Chromite-mesocumulate would be an acceptable alternative term. Oxide-rich cumulates are variously mined as a source of chromium, titanium, vanadium (occurring in magnetite) and platinum group metals (PGMs). Sulphides, accumulating typically as immiscible liquid drops due to oversaturation of sulphur within the magma, can occur in large, ore-grade concentrations. Key metals that are mined from sulphidic cumulate deposits are nickel, copper, cobalt, the PGMs and gold, with sulphides like pyrrhotite (FeS), pentlandite ((Fe,Ni)$_9$S$_8$) and chalcopyrite (CuFeS$_2$) dominating the ore assemblages, accompanied by a wide range of accessory species. The PGMs may occur in the native state, as sulphides or arsenides and as alloys.

In the UK, cumulate bodies have been worked for chromite in the Shetland Islands: they are locally enriched in PGMs but not, so far as we currently know, to an economic extent. Other cumulates have been prospected for nickel and PGMs in parts of Aberdeenshire. However, elsewhere in the world, such mineralization is developed to a spectacular degree. A prime example is the 'Merensky Reef', a layered cumulate belonging to the Bushveld Igneous Complex of South Africa, which is the most important source of PGMs in the world, with copper and nickel as by-products.

3.2.2 Pegmatites

Pegmatites may form as late-stage products of cooling igneous intrusions and also, and often more widely, as a product of partial melting of rocks during high-grade regional metamorphism. They represent a bridge between, on the one side, purely magmatic deposits and, on the other, hydrothermal deposits. Pegmatite magmas always have a very high concentration of dissolved water, and in some cases the aqueous fluid is exsolved from the melt; in others deposition is by aqueous fluid only, with no silicate melt present.

Typical pegmatites are vein-like bodies composed of very coarse quartz, feldspar and mica (Fig. 3.1): indeed, some are mined for sheet mica that may occur as 'books' several metres in size. However, the range of minerals

Figure 3.1 Pegmatite
Cut hand-specimen of pegmatite from eastern Scotland, consisting of very coarse-grained crystals of feldspar, quartz and dark-coloured tourmaline.

that pegmatites may contain is considerable, and many examples are enriched in various rare elements. The reason for this concentration of rare elements in such late-stage deposits is that they are chemically incompatible with the minerals that crystallize from the magma during the main stage of solidification of an intrusion. Thus, as the minerals crystallize and the rock solidifies, any residual silicate melt progressively becomes more enriched in whatever incompatible elements were originally present. These include metals like lithium, beryllium, caesium, tin, niobium, tantalum and tungsten, metalloids like boron and non-metals like fluorine. Commercially important pegmatite minerals include spodumene (LiAl(SiO$_3$)$_2$), beryl (Be$_3$Al$_2$(SiO$_3$)$_6$), cassiterite (SnO$_2$), columbite ((Fe, Mn) Nb$_2$O$_6$) and tantalite ((Fe, Mn)Ta$_2$O$_6$). In addition, some pegmatite minerals are of importance as gemstones, such as beryl, or as mineral specimens, such as tourmaline (generic formula (Ca,K,Na) (Al,Fe,Li,Mg,Mn)$_3$(Al,Cr, Fe,V)$_6$(BO$_3$)$_3$(Si,Al,B)$_6$O$_{18}$(OH,F)$_4$). In some parts of the world, pegmatites are worked commercially for tourmaline specimens that retail for tens or even hundreds of thousands of dollars. In the UK, pegmatites are particularly common as intrusions in the Precambrian Moine Group metasediments of NW Scotland, where they have been worked in the past for their large mica crystals.

3.3 Hydrothermal mineral deposits

Hydrothermal literally means of, or relating to, hot water. Hydrothermal mineral deposits, deposited from hot aqueous solutions, constitute the most diverse class of all mineral deposits, with a seemingly unlimited variety of possible settings and mechanisms of formation. They can occur in association with intrusion and eruption of igneous rocks, during the development of sedimentary basins and during regional, contact and dynamic metamorphism. This is less of a surprise when one considers the abundance of water on the Earth's surface and in the crust, and the plethora of available sources of heat.

Hydrothermal mineral deposits vary in size and complexity from simple, outcrop-scale quartz veins through to complex polymetallic deposits forming **lodes**, disseminations or **stockworks** of veinlets shot through large bodies of rock, replacement-bodies occurring in reactive host-rocks (such as limestones) and the products of seafloor hydrothermal fluid **exhalations** forming sediment-like accumulations.

Igneous activity and hydrothermal activity go hand-in hand in almost all instances. The heat-engine of a magma chamber is at work both during the molten phase and after solidification, as the rock slowly cools down. As such, it acts as a source of convective energy; groundwaters, making their way down through the strata by gravity, are heated in the vicinity of the chamber to very high temperatures and driven back up in the direction of the surface via whatever fracture-systems are available to them. They may also become mixed with waters of magmatic origin, driven off from the magma chamber during the final stages of solidification. Under such physical conditions, the heated waters are able to leach and dissolve a wide range of elements from the strata they pass back through: they become a chemical soup charged with cations and anions of metals and nonmetals alike. The **leaching** of the strata may also be accompanied by the precipitation of new minerals, thus changing the bulk mineralogy of the rocks that the fluids have moved through. Such changes are typically most marked along the walls of the conduit fractures, hence the term '**wall-rock alteration**'; alteration assemblages can be diagnostic for certain types of important mineralization and are thus mapped in the field during mineral exploration projects (Fig. 3.2).

Alteration can take many forms and is to a large extent controlled by the reactivity and permeability of the rocks that the fluids are passing through. Thus, a porous sandstone may become cemented by silica into a much harder rock: a limestone may be completely replaced by other minerals or the mafic minerals and feldspars of a basic intrusion may be replaced by other minerals.

Figure 3.2 Wall-rock-alteration
Wall-rock alteration: a matrix of dark psammite (metamorphosed coarse sedimentary rock) is cut by a thin quartz-carbonate-sulphide vein just a couple of millimetres in thickness, but a halo of altered (carbonatized and sericitized) rock extends out from either side of the vein for much greater distances, forming the buff-coloured zones.

Common alteration patterns include **silicification** – the partial replacement of a rock by quartz; **carbonatization** – involving replacement by calcite and/or dolomite; **sericitization** or **chloritization** – the development of the sheet-silicates sericite and chlorite respectively, and **pyritization** or **arsenopyritization**, in which wall-rocks become impregnated with pyrite or arsenopyrite. In many instances an alteration assemblage may consist of several of these and other minerals: thus a dolerite sill may lose its original feldspars, pyroxenes and amphiboles and instead now consist of a mixture of calcite, sericite and chlorite.

When hydrothermal fluids pass into open fractures, they typically deposit minerals along the fracture walls; thus the fracture becomes a mineral vein or lode. The minerals partially or completely fill a fracture and they may be coarsely or finely crystalline, depending on how slowly they precipitate from solution and, consequently, how much time the crystals have had to grow. Precipitation of minerals from solution is triggered by changes in the prevailing conditions in which the hydrothermal fluid exists; for example, if a fracture suddenly dilates, the confining pressure regime will sharply drop off. Ions that are stable in solution in highly pressurized fluids may then become less so, the result being that they partially or totally precipitate out, combining with other ions to form minerals. Cooling within a hydrothermal fluid will likewise affect the solubility of various ions; two fluids of differing composition that meet and mix can trigger precipitation, as can a fluid encountering reactive wall-rocks.

Fractures that become mineralized may be pre-existing, may be forming at the same time due to extensional tectonic processes, or may actually be formed or propagated via the **hydraulic** action of the fluids themselves. The latter process is important in lode development and is worthy of a brief description. Consider a highly pressurized liquid (be it a magma or hydrothermal fluid) trapped along a fracture-plane. It is held where it is by the confining pressure of the walls of its host fracture, these being a function of the **lithostatic pressure** or load caused by the overlying strata. However, if that fluid becomes so pressurized that its fluid pressure exceeds the confining pressure, it automatically overcomes the tensile strength of the rock, so that fracture walls are forced apart and the fracture-plane extends outwards. As this dilation occurs, the fluid pressure-drop may be so sudden that it becomes much less than the pore-water pressure in the wall-rocks, which then explode out into the fracture, leaving a mass of rock-fragments – a **breccia** – cemented by the newly-precipitated minerals. The system thus re-seals itself until the next pulse of fluid migrates into the fracture and the process repeats itself. It is not surprising the damage that a small amount of pressurized hydraulic fluid can do when one considers that a mechanic's trolley-jack can easily lift a vehicle weighing over a ton by the action of a piston in a cylinder containing a few tens of millilitres of hydraulic fluid (see Fig. 3.3A, B.

Figure 3.3 Examples of hydraulic breccias
Polished slabs of brecciated siltstone cemented by quartz and sulphide minerals. In the first image (**A**), the siltstone fragments can in some cases be visually reassembled, as the sedimentary banding can be traced from piece to piece. Such shattering and rapid recementation by minerals is a common feature of hydraulic vein-breccias; however, other examples, such as the second image (**B**), reveal signs that some rock clasts have travelled along in the hydrothermal fluid, with collisions between clasts leading to their abrasion and rounding – as well displayed in the centre R of the image, implying a delay between brecciation and mineral precipitation. This breccia is cemented by clear quartz and brown sphalerite.

Hydrothermal fluids generated in an igneous regime can, and often do, make it all the way to the surface, where they emerge as **fumaroles** and geysers. In such cases, the superheated water explosively bursts forth and, with the accompanying and sudden confining pressure-drop and cooling, the waters deposit their dissolved contents as relatively low-temperature mineral assemblages. Indeed, within igneous hydrothermal systems there is commonly a marked vertical zonation, with high-temperature (**hypothermal**) mineral assemblages (for example, tin- and tungsten-rich) in and just above an intrusion, medium-temperature (**mesothermal**) assemblages (copper, zinc, lead, arsenic, gold) at moderate depths and low-temperature, shallow to surface-formed (**epithermal**) assemblages right at the top of the hydrothermal plumbing system, often containing gold, silver, arsenic, mercury and so on. In the field, it may be the case that only parts of such a system remain in place, the upper zones having been removed by erosion. Hence, in parts of the important, granite-related tin-producing region of SW England, tin-tungsten lodes are exposed at surface; in other areas copper ores prevail at surface with tin and tungsten occurring at depth.

In areas undergoing submarine igneous activity, the processes are broadly similar, but with the important additional factor of relatively hot fluids reaching the surface – the sea-bed – where they spew forth as the so-called '**black smokers**'. The dark colour is due to fine-grained metal sulphide precipitates that form as a consequence of changes in pressure

and temperature plus reaction with the surrounding seawater. Such precipitates, being heavy, sink to the sea-bed where, if activity is intense and prolonged, they may accumulate in significant quantities to form, in time, massive sulphide deposits, often with sedimentary textures. In deposits of this type, if an igneous connection can be clearly demonstrated, the term **Volcanogenic Massive Sulphide** (typically expressed as the acronym VMS) is used. A classic UK example of a VMS deposit, once the largest copper mine in Europe, is at Parys Mountain on Anglesey, worked for over two centuries and under active exploration at present, the target being not only copper but the accompanying lead, zinc, silver and gold (see Figs 3.4, 3.5).

Within some igneous intrusions, mineralization occurs not only as lodes, but also as systems of veinlets known as stockworks or disseminations through the body of the rock (**porphyry-type mineralization**). In such deposits, tonnages may be vast but the overall concentration of metals low. Porphyry-type mineralization relies in its development upon the internal shattering of a crystallizing igneous intrusion on a microscopic scale, caused by the boiling of volatiles, most importantly water; the fracturing is caused by the volume-increase that accompanies the transition from liquid to gas. Such shattering creates innumerable pathways into which fluids then make their way and deposit minerals. Porphyry-type deposits typically occur in the upper parts of

Figure 3.4 Parys Mountain
View of the Great Opencast at Parys Mountain, once Europe's largest copper-mine. Worked principally in the 18th and 19th centuries, it exploited a volcanogenic massive sulphide deposit that also contained lead, zinc, silver, gold and much pyrite, hosted by Lower Palaeozoic sedimentary and volcanic rocks.

Figure 3.5 Volcanogenic Massive Sulphide mineralisation, Parys Mountain
An example of the 'bluestone' ore at Parys Mountain, with faint sedimentary banding visible. The sulphides were precipitated from solution in seawater and accumulated as chemical sediments on the sea-floor. Chalcopyrite, sphalerite, pyrite and galena are all present in this block, and polished sections have revealed the presence of a wide range of uncommon minor sulphides.

intrusions and in the roof-rocks immediately above, and can be important as sources of copper, molybdenum, tungsten and gold. They are typically enveloped by concentric zones of wall-rock alteration – thus many such deposits have a so-called 'pyrite halo', deposited as the fluids make their way out into the surrounding rocks, cooling as they go. Because volatile-rich magmas are required for the formation of such deposits, most porphyry-type mineralization is associated with acidic to intermediate igneous rocks, such as granites and diorites. A well-known UK example of this style of mineralization – complete with pyrite halo – is the unworked

Coed y Brenin copper deposit in North Wales. Nearby, at high structural levels above the ore-zone, there occur pipe-like bodies consisting of shattered and silicified wall-rock cemented by banded quartz and pyrite, with enrichment in arsenic, antimony, silver and gold, with a clearly epithermal character.

Sedimentary processes may also lead to the set-up of mineralizing hydrothermal systems. A key factor here is the crustal subsidence that accompanies the development of a **sedimentary basin**. When sediments accumulate rapidly in a subsiding sedimentary basin, individual layers find themselves becoming buried to

progressively greater depths and subjected to increasingly greater pressures and temperatures. At some point during this progression, a layer of sediment will become converted to solid rock in the process termed **lithification**. Seawater, trapped within the sediment, will then to a large extent be expelled: such a fluid is termed a **connate brine**, to distinguish it from hydrothermal fluid of magmatic origin, or groundwaters originally derived from precipitation (rainwater), the latter being known, appropriately, as **meteoric waters**.

As a consequence of the chemical reactions during lithification, connate brines carry metal ions. The loading is increased as they pass along pathways up and out of the basin, driven by the high **geothermal gradient** that exists between the surface and the depths of a typical sedimentary basin. The elements they become enriched in vary according to the rocks that they travel through as they are driven up out of a basin, traveling considerable distances (>100 km) in some cases. They may work their way into existing fault-systems or, by hydraulic fracturing, open up new faults; again, as temperatures and pressures change during their journey up through the crust, minerals are deposited along fracture-walls, forming lodes. Reactive rocks, such as limestones, may be extensively replaced, via the process known as **metasomatism**, by minerals that extend far outwards laterally from the fracture walls, forming mineralized structures traditionally referred to as 'flats'. The fluids may make it to the surface, where, exhaling onto the seabed, they flow along its floor, collecting in topographical lows, to form '**sedimentary-exhalative**' deposits (Fig. 3.6).

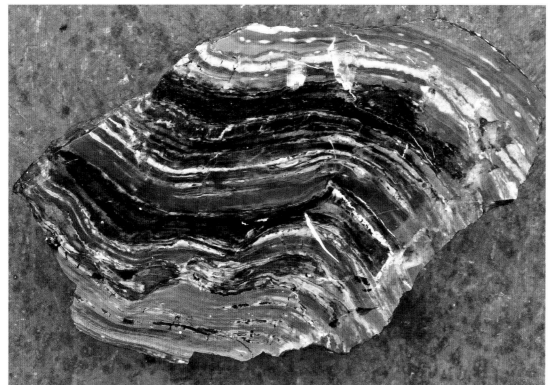

Figure 3.6 Sedimentary manganese-ore
Sedimentary-exhalative mineralization: a polished section, 20 x 15 cm in size, through part of the strikingly coloured Manganese Ore Bed of the Harlech Dome, North Wales. The bed, largely consisting of fine-grained manganese-bearing carbonate and silicate minerals, is an important marker horizon in the local Cambrian succession, and old mine-workings are to be seen wherever it crops out.

Deposits formed by fluids expelled from sedimentary basins are important sources of metals such as lead, zinc, copper, iron, manganese and industrial minerals like baryte and fluorite. Many of the UK's base-metal mining districts exploited ore-bodies formed from connate brines, and where the host-rock is limestone, as in the Pennine orefields, metasomatic flats can be common and very extensive. In the Irish Midlands, sedimentary-exhalative lead-zinc-baryte deposits occur on a large scale and have been developed and mined over recent decades. Finally, some mineral deposits formed by these processes are highly enriched in other less common elements; in the base-metal lodes of Central Wales, for example, enrichments of silver, nickel and cobalt are present.

Beyond diagenesis and lithification, further progressive burial of sedimentary and igneous rocks leads to more changes in mineralogy in the process known as burial-related metamorphism. This can lead to the dehydration of existing minerals, and thereby to the generation of further hydrothermal fluids. Again, these migrate through pathways in the crust, traveling along fractures, altering the wall-rocks, leaching elements and redepositing them in lodes. At the depths – and thereby the pressures and temperatures prevailing – such lodes tend to carry mineral assemblages of a mesothermal nature. An important UK example is the Dolgellau gold-belt of North Wales, in which numerous quartz lodes hosted by Cambrian sedimentary and igneous rocks carry concentrations of copper, lead, zinc, iron, cobalt, arsenic, bismuth, tellurium, silver and gold.

An interesting variety of hydrothermal mineralization can occur on a localized scale during regional dynamic metamorphism. Affecting competent rock units within ductile matrices, such as sills hosted by thick mudstone sequences, mineralization occurs during compressive deformation and cleavage development. While mudstones deform by plastic means, brittle lithologies deform by fracturing; the resultant pull-apart structures basically constitute low-pressure zones in an overall high-pressure regime. Any fluids in the immediate vicinity will therefore migrate to such areas and precipitate their contents on the fracture-walls. Such 'Alpine Fissure-type' veins are common in the more deformed sequences in the UK, but are only important as a source of mineral specimens, though it has been suggested that, because they contain minerals that have potential for isotopic dating, they may yield useful data with regard to the timing of deformation. In the Alps, however, they have been mined over the centuries not only for specimens but also for optical-grade clear quartz from the sometimes enormous crystals that they contain (see Figs 3.7 and 3.8).

Figure 3.7 Alpine-type veins
Localized hydrothermal mineralization in the form of quartz-dominated veins intersecting a tuff-turbidite bed (pale and banded) in an otherwise mudstone-dominated sequence of Ordovician age in North Wales. Compressive deformation later converted the mudstone to slate; the slaty cleavage, discordant to bedding, can be seen dipping at 45° from R to L. The response of the relatively brittle tuff-turbidite bed to the compressive deformation was extension along-strike and brittle fracturing at steep angles to bedding. Each fracture was then filled by quartz and other minerals derived from very localized hydrothermal systems, so that the minerals contain elements remobilized very locally.

Figure 3.8 Synchysite
A small (35 x 20 mm) specimen collected from one of the veins in Fig. 3.7 cutting the tuff-turbidite. The assemblage consists of fine-grained quartz, on top of which are preudohexagonal crystals of synchysite-(Ce), a rare-earth bearing mineral with the formula $Ca(Ce,La)(CO_3)_2F$ but containing other lanthanides. The black bipyramidal crystals are the titanium dioxide mineral, anatase.

Finally, an important class of hydro-thermal mineral deposits can occur in the contact zone of intrusions where the host-rock is of a chemically reactive nature, such as limestone. A wide range of chemical reactions can occur in such situations, especially where the intrusion is volatile-rich: the hydro-thermal fluids generated metasoma-tize the host-rock, replacing it with mineral assemblages that reflect both the chemistry of the intrusion and of the replaced host. Limestones replaced by calcium silicates and other minerals are referred to as '**skarns**': they typically consist of pyroxene, garnet, actinolite, magnetite and epidote, but depending on the volatiles present, they may be rich in tourmaline, topaz, beryl, fluo-rite and apatite plus a range of metals including tin, tungsten, copper and gold. Copper and tin were formerly obtained from skarn-type deposits in the Meldon area of Devon, in the contact-metamorphic aureole of the Dartmoor Granite.

To ask what minerals typically occur in hydrothermal mineral assemblages is to ask the wrong question! This class of mineral deposits is characterized by an enormous diversity in elements and minerals. Thus there are native metals, oxides, silicates, carbonates, sulphates, halides, sulphides, selenides, tellurides, arsenides and antimon-ides occurring variably from deposit to deposit, some in abundance and others only present as tiny, microscopic inclusions in other minerals. Some metals are typically associated with one another: examples include lead and zinc (as galena, PbS, and sphal-erite, ZnS); lead and silver, the silver occurring in solution in the galena or

as inclusions of argentiferous minerals like tetrahedrite $(Cu,Fe,Ag,Zn)_{12}Sb_4S_{13}$); silver with nickel, cobalt, arsenic and bismuth (the so-called 'five metals association'); pyrite (FeS_2) with arsenopyrite (FeAsS), and so on.

3.4 Chemically-precipitated atypical mineral concentrations

Limestones and evaporites have already been discussed under common minerals of sedimentary origin. However, some other pre-cipitates are much more restricted in occurrence, either spatially or in terms of geological time. A classic example of an unusual, chemically precipitated class of mineral depos-its is **banded iron formation** (BIF).

BIFs are sedimentary rocks that typi-cally consist of alternating thin layers of haematite, magnetite and silica, associated with mudstones and cherts of marine origin. An important source of iron, they are generally restricted in occurrence to Precambrian strata, with the most extensive occurring in sequences dating from the early Pro-terozoic, between about 2.5 and 1.8 billion years ago. This was a critical point along the timeline of the evolu-tion of Earth, for it was marked by the coming into abundance of marine-dwelling photosynthetic cyanobacteria and, as a consequence, a significant rise in atmospheric oxygen levels, including what has come to be known as the '**Great Oxygenation Event**', about 2.4 billion years ago.

Today, iron is readily oxidized at the Earth's surface to various stable, rusty-looking hydroxides. Back in the depths of the Precambrian, prior to the Great Oxygenation Event, this

was not the case: Earth's atmosphere contained little or no free oxygen. In such circumstances, the weathering of iron-bearing minerals like pyrite would have led to the iron becoming mobile in drainage systems and hence occurring in abundance, dissolved in the oceans. It was the interaction between this mobile iron reserve and evolving life that is thought to have led to the development of the BIFs, and specifically the proliferation of oceanic photosynthetic cyanobacteria in the near-surface waters. The oxygen pro-duced by these algae reacted with iron cations in the water to form iron oxide particles, which sank and accumulated on the sea-floor, thereby reducing iron concentrations in the water. However, **algal blooms**, just as today, tend to use up all their available nutrients or alter their environmental chemistry; in consequence, a mass die-off would periodically occur and the remains of the algae would sink to form a band of fine-grained siliceous sediment, before the ecosystem recovered and iron oxides would again begin to accumu-late. In this manner, a sequence built up, typically consisting of alternating iron oxide (finely laminated haematite (Fe_2O_3) and magnetite (Fe_3O_4)) and silica-rich bands. Just what timespan these bands represent is not entirely clear. That there is a cyclicity present is readily apparent, but whether it is sea-sonal or related to some other cycle is not well understood.

Once the atmosphere became well oxygenated, the availability of soluble iron decreased, although some would still have been produced by decay of iron-bearing sulphides such as pyrite, and under-sea hydrothermal fluid exhalation would have provided still more. But the more the atmosphere became oxygenated, the more the oceans became oxygenated too, reducing the dissolved iron content of the oceans markedly. Banded iron formations less than 1.8 billion years in age are relatively uncommon: some are associated with the late Neoproterozoic Cryogenian System (850–635 million years ago), a period during which very extensive and possibly global glaciations (the so-called '**Snowball Earth**' events) occurred. The presence of BIFs in rocks dating from this time have been suggested to indicate that such extensive icing-over of the oceans occurred that interaction between sea-water and the oxygenated atmosphere was prevented. However, there remain many unanswered questions regarding the Cryogenian glaciations and the environmental chemistry at the time.

3.5 Weathering-related mineral deposits

Weathering is a familiar process to all of us. It involves the chemical reactions between chemical compounds in the atmosphere and chemical compounds on the planet's surface. When your car's exhaust pipe falls apart noisily, it is because the iron making up the steel from which it was constructed has, over several years, reacted with oxygen and rainwater to form rust. But that's a relatively fast example involving a relatively unstable compound. The compounds making up the vast majority of Earth's land surface – those that make up rocks – are by contrast very slow to react. As a consequence, large-scale weathering is a process that takes place on a timescale of millions of years.

The weathering process begins when carbon dioxide dissolves in rainwater. The resulting solution is weakly acidic:

$$CO_2 + H_2O = H_2CO_3 \text{ (or carbonic acid}$$
– the old name for carbon dioxide was carbonic acid gas)

Rainwater containing carbonic acid is able to react with most minerals at varying rates according to their chemical stability. Now, some naturally-occurring minerals are very stable. Think about gold, eroded mechanically from ore-deposits and then recovered maybe many thousands of years later from rivers by panning. At the other end of the scale are very unstable minerals, such as most sulphides. The common iron sulphide, marcasite, can be so unstable that it reacts with the slightest bit of moisture in the air and starts to decay, providing a headache for museum mineral collection curators.

Although the weathering of rocks is a slow process in most cases, in some instances it may be accelerated. For example, pyrite may occur in some rock-types as scattered crystals or pyritized fossils. As weathering sets in, pyrite will react more rapidly with rainwater due to its instability, and when it breaks down, it liberates sulphuric acid. A typical equation for the reaction is as follows:

$$2FeS_2 + 6H_2O + 7O_2 = 2Fe(OH)_2 + 4H_2SO_4$$

iron sulphide + water + oxygen = iron hydroxide + sulphuric acid

The sulphuric acid then sets about reacting with other adjacent minerals, being a more powerful solvent than carbonic acid. For this reason, old mines can pose real problems in environmental terms: because most ore-deposits contain pyrite, the exposure of fresh sulphides in mine-workings leads to the occurrence of the reaction detailed above. The sulphuric acid then attacks other metal-liferous minerals, thereby liberating metals into solution, leading to what is termed acid mine drainage – acidic water containing a toxic cocktail of metals, often with a bright red or yellow colour due to suspended iron-ochres.

In completely natural systems, ore-deposits weather via similar mechanisms but at a slower pace, the end-products being relatively stable **secondary minerals**. A marked vertical zonation may be present in deep-weathered ore-deposits. At the surface, there remain only the most stable **primary minerals** – such as quartz – and the most insoluble secondary minerals, such as iron oxides. Together these form a rusty-coloured cap on the rest of the ore-deposit, known as a '**gossan**'. An old mining adage goes thus: 'there is no such lode as that, which wears an iron hat'. However, the presence of an iron-gossan may in reality just mean that primary pyrite is present at depth. Conversely, a gold-bearing pyrite lode may have a gossan in which the gold particles remain yet other primary ore-minerals have gone. Due to this removal of other metals, the gold grade of the gossan may be much higher than that of the primary mineralization below, sometimes to such an extent that it is only the gossan that is worth mining in economic terms. Whatever the case, prospectors will not overlook any gossans, always sampling them and having them analysed for gold (Fig. 3.9).

With increasing depth, weathered ore-deposits feature a wide range of secondary minerals of the metals

Figure 3.9 Gossan
A classic example of a gossan: a matrix consisting mostly of the iron oxide, goethite, plus some residual quartz and flecks of the lead sulphate, anglesite, from Parys Mountain in North Wales. This is the oxidised equivalent of the "bluestone" ore, illustrated earlier in this chapter.

present: carbonates, phosphates, arsenates, vanadates and so on. Both cations and anions may be mostly derived from the weathered primary mineralization, or in some cases anions such as phosphate and carbonate may be derived from phosphatic or carbonate host-rocks. The secondary minerals that are formed can occur in sufficient quantities to be worked as ores, and important ores at that, because in some cases secondary minerals are richer in the key element than their primary predecessors. This is especially true in the case of copper. Some secondary minerals have considerable value as ornamental materials, such as the copper

carbonate, malachite; well-crystallized and often colourful secondary minerals also form a large part of the multimillion dollar mineral specimen market (Fig. 3.10).

The richest secondary mineralization tends to occur around the water-table where the chemical environment is most reducing in nature. In this zone, strongly-enriched metal-occurrences are met with: for example, copper is reprecipitated as the native metal, the oxide cuprite (Cu_2O, 88% Cu) and the enriched sulphides like chalcocite (Cu_2S, 79.8% Cu). Compared with the copper content of the chief primary copper mineral chalcopyrite ($CuFeS_2$,

34.63% Cu), it becomes obvious that such enriched ores are of high value.

Weathering of certain rocks can produce important economic mineral deposits, one example being granite, whose feldspars weather to produce kaolinite ($Al_2Si_2O_5(OH)_4$) or China Clay. Dissolved carbon dioxide is one agent in the process; hydrogen ions in groundwaters are also important, as the chemical reactions include hydrolysis, involving the reaction of water with feldspar to produce clay minerals and soluble ions:

$$2KAlSi_3O_8(s) + 2H_2CO_3(aq) + 9H_2O(l)$$
$$= Al_2Si_2O_5(OH)_4(s) + 4H_4SiO_4(aq) +$$
$$2K+(aq) + 2HCO_3–(aq)$$

orthoclase + carbonic acid + water = kaolinite + silicic acid, potassium and bicarbonate ions in solution

Via reactions such as the above, an entire body of granite may, over millions of years, be reduced to a loose sand of quartz and mica set in the kaolinite clay. Such clay-rich soft rocks are easily worked and the clay, which has many industrial uses, is extracted from its matrix by washing with water.

The intensive weathering of olivine-rich ultrabasic intrusive rocks such as peridotite can lead to the concentration of nickel, a metal that is often present in such rocks in low concentrations. Such deep weathering takes place in tropical conditions – the warmth encourages the chemical reactions. The products that are formed in these '**nickel laterites**' are nickel-bearing secondary oxides and silicates. Nickel concentrations can be enriched from, say, 0.3% in unweathered rock up to 2% or more, and with vast tonnages available in the larger weathered intrusions, which are worked from the surface down towards

Figure 3.10 Azurite
From an oxidation zone at a copper mine in Anhui Province, China. Rather earthy malachite and iron oxides have replaced primary ore (probably chalcopyrite) and have then been overgrown by crystalline azurite followed by a second generation of relatively pure malachite, suggesting that geochemical conditions were fluctuating during the process – malachite is stable under a wider pH range than azurite.

the limit of weathering, nickel laterites are an important source of the metal.

3.6 Placer mineral-deposits

Weathering of rocks and ore deposits is one important way in which some minerals become richly concentrated; another is erosion. This is the physical process by which rocks are mechanically worn down from the splitting action of frost, sometimes bringing whole crags collapsing downwards, to the fine abrasion of sand-grains as they are tumbled along a river-bed,

continually bumping into one another. These processes transport rock fragments into new environments where they are reduced in size and eventually redeposited. The transportation involves forces determined by environmental constraints: rainfall patterns and a stream's gradient both strongly influence the velocity of the flow. It's all about energy: the higher the energy, the bigger and heavier the objects that may be moved along.

Placer mineral deposits typically consist of concentrations of resilient

minerals (both in terms of their weathering- and abrasion-resistance) which, because of their relatively high density, either resist being carried along by the flow while everything else is removed (lag-deposits) or are preferentially redeposited where the energy in the transport system suddenly drops off (for example, where a steep river-course flows into a flat-lying section). The transport system may be rivers, as mentioned already, or tidal currents – there are beach-placers – or less commonly, wind. These mechanisms of erosion and transportation/concentration have been present since Earth developed a solid crust with free water available on it; therefore ancient sediments may likewise contain concentrations of heavy minerals.

Many minerals are won from placers: examples include cassiterite (SnO_2), '**black sand**', typically dominated by magnetite and ilmenite ($FeTiO_3$) – a source of titanium, tantalite ($(Fe, Mn)Ta_2O_6$) – the main source of the metal tantalum and various rare earth minerals. However, gold is the best-known placer mineral, obtained from rivers since time immemorial, so it forms a useful case-study. The following comments apply to the other placer-minerals too.

For a river to be gold-bearing, an original source for the metal is a prerequisite: this may be one or more gold-bearing ore-deposits or an ancient gold-bearing **palaeoplacer** such as a conglomerate that is being reworked by erosion within the catchment. Geologists exploring a region for gold will take panned heavy mineral concentrates from the river-sediments and visually examine them for the presence of gold grains. If panned concentrates consistently reveal alluvial gold grains, it follows that there should be a source for that gold somewhere in the catchment. However, in glaciated terrains, the erosion of glaciated till deposits may liberate and concentrate gold in the rivers, a process that can provide a real headache to exploration geologists. Glaciers, and especially ice-sheets such as those that occurred in the UK during the Quaternary, can transport huge quantities of eroded rock debris over hundreds of kilometres. Subsequent reworking of the debris by rivers can lead to gold being present in the stream-sediments, when it does not in fact occur anywhere in the bedrock of the catchment! Such situations are referred to as '**displaced anomalies**'– a few grains of gold in a pan of river sediment would be regarded as anomalous, but in this case the source of that anomaly is far, far away.

Because of its high density (19.3 g/cm^3, compared to quartz which is only 2.65 g/cm^3), gold requires an energetic environment to be transported, and this especially applies to larger particles such as flakes and nuggets, which may only be moved during the worst flash-floods. Fine-grained gold – scales and dust – is more readily moved but tends to drop out of suspension at the slightest drop in energy or following contact with a roughly-textured substrate – a 'gold-trap'. Experienced prospectors use this knowledge in their search: they will wash moss that is inundated during high river-levels for 'flood-gold' and they will wash gravels along low-velocity sections of rivers where deposition is likely to have occurred, the inside curves of bends being one prime spot to look. The coarser gold,

once it ends up in the river system, will tend to find its way vertically downwards through the riverbed sediment until it reaches bedrock, where it will become lodged in cracks. In some rivers, prospectors don diving gear, and using special tools they will work along the underwater bedrock, picking flakes of gold out of the cracks. In others, powered suction-dredges are deployed to suck up the entire riverbed and run it over a **sluice** – a trough floored with wooden riffles set at right angles to the flow. Imitating bedrock cracks, these structures trap the heavy minerals, including the gold; the flow of the water takes the lighter sediment off back into the river (see Figs 3.11 and 3.12).

Palaeoplacers are important sources of gold and other minerals, the best-known example being the auriferous comglomerates of the Witwatersrand in South Africa, which are the remains of an ancient (late Archaean to early Proterozoic) delta system in which gold, eroded from neighbouring older rocks, was deposited. Other heavy minerals concentrated in clastic sedimentary rocks may have important potential in research: zircon, for example, may be isotopically dated by the uranium–lead technique, thereby establishing the age(s) of the source rocks that were eroded to make up the sediment.

Figure 3.11 Heavy concentrate

A concentrate of heavy minerals from a stream-bed in central Scotland, dominated by garnets, haematite, ilmenite and magnetite but also including cinnabar (brick-red grains) and gold (a large grain is visible bottom-centre). This was the result of putting several bucketfulls of stream sediment through a sluice-box to wash away the lighter material.

Figure 3.12 Gold

A collection of gold nuggets up to 12 mm in size from the same central Scotland location as the heavy concentrate. The one second from top left is intergrown with quartz. Coarse gold like this is readily seen and picked from a sluice, although the collection represents many days of prospecting!

4 Mineral collecting: where science and leisure overlap

4.1 Introduction

The pursuit of mineral collecting has a long pedigree stretching back several centuries and pre-dating modern academic mineralogy by a considerable degree. The early mineralogists were frequently gentleman scientists with the time available to procure and investigate mineral specimens, and today many new mineral occurrences, or new minerals altogether, have come to light as a consequence of the activities of non-academic collectors. By non-academic I mean people who are not necessarily formally qualified in the subject, but who nevertheless have a considerable comprehension of mineralogy, and for whom the term 'amateur' seems inappropriate, as their depth of knowledge can equal that of any university researcher. On top of that, there are the commercial collectors and dealers who vary from the highly knowledgeable through to bulk commodity brokers (for example, beds of amethyst crystals from South America are imported in their tonnes) and those who simply like to spend a day out with a few friends in the fresh air, and are satisfied with a few nice quartz crystals. It's a big field with a wide range of individual perspectives and priorities; research, aesthetics, money and pure enjoyment all play their part.

In the UK, mineral collecting really took off in the eighteenth and nineteenth centuries: although the number of participants was relatively small compared to the present day, the material preserved was of exceptional scientific importance. These people had a major advantage in that there were far, far more working mines back then, access tended to be a lot easier, and access problems were more readily circumvented. A classic example was that of the Cornish collector and dealer Richard Talling: banned from the Herodsfoot mine near Liskeard, which had produced what are still regarded as world-class specimens of tetrahedrite and bournonite ($CuPbSbS_3$), he purchased a block of shares in the venture as a workaround and was able to resume his activities!

Specimens from the mines of Cornwall, Devon, Derbyshire, Cumbria, the North Pennines and the Leadhills district of Scotland that were collected at this time rival, in aesthetic terms, those from anywhere in the world, and in some cases are regarded as the finest known examples of their species. During the twentieth century, economic factors saw the closure of many productive localities, although some districts such as the North Pennines saw active mining (and specimen production) continue. With the massive expansion of the aggregates industry, collectors turned to working quarries as potential sources of specimens and, although access can be difficult to arrange due to safety legislation, important finds continue to be made to the present day. Another switch in emphasis concerns specimen size: whilst the early collectors were firmly focused on impressive **hand-specimens** and larger, museum-sized pieces, the relative lack of such material in more recent times has led, quite literally, to a zooming-in. The advent of affordable binocular microscopes led to the emergence of '**micromounting**', the study of millimetre-sized and smaller crystals occurring on specimens smaller than your average thumbnail. Thus, collectors turned their attention to old mine-tips, digging through the spoil and taking likely material home to be broken up, and the specimen-grade pieces mounted for viewing in clear plastic boxes, which also serve, importantly, to keep the dust out. Many new occurrences of rare minerals and discoveries of minerals previously unknown to science have stemmed from the activities of micromounters (Fig 4.1A, B).

4.2 Mineral collecting: sites and safe collecting

All sorts of sites have the potential to yield interesting and/or attractive mineral specimens. A scan through

Figure 4.1 Micromount and Elyite

(**A**) A microcrystalline specimen kept in a plastic box. (**B**) Detail of the same specimen, showing a delicate group of microcrystals just 1–2 mm in length. The mineral is the very rare basic lead-copper sulphate, elyite.

the literature shows that, in the UK alone in just the past 30 years, important finds have come from working and disused mines, stone quarries, road cuttings, beaches and sea cliffs, river alluvium, river bedrock and outcrops in pastures and on hillsides. Such a range of potential sites does not, however, mean that finding good material is easy, but collectors can improve their own odds to an extent.

There are several ways to go about this. For the individual prepared to go it alone, the first step – and we'll use a UK-based would-be collector as an example – is to pick an area and read up on its geology, with the British Geological Survey's Regional Geology Guides series being an ideal starting-place, as they give mineralization some coverage. More detailed descriptions of mineralization are given in the geological memoirs that accompany many of the 1:50,000 geological maps. Via these publications, an idea of the potential scope of an area may be gained. Further research would include the mineralogical literature and local industrial archaeology publications. In recent decades, the internet has made it far easier to track down relevant publications. By this point, the collector should have an idea of what locations, both known and potential, are available to visit.

The next step involves determining whether access is possible. Most sites are situated on private land, and access is entirely at the discretion of whoever owns that land, so it is essential to find out who that is and contact them. Asking around locally usually works: however, permission is not guaranteed. Working mines and quarries can be hard to access due to safety legislation; although many quarry managers are quite happy to show visitors around by prior arrangement, that may not offer much opportunity to engage in serious collecting. Some sites may be protected as **Sites of Special Scientific Interest (SSSIs)**, in which case additional permission may be required from the relevant government agency. There can, therefore, be a lot of hoops to jump through!

There is an easier and gentler route into mineral collecting – join a club. In the UK, there are a number of local clubs and national groups such as the Russell Society, with several local branches. The advantages are that you will be able to tap into decades' worth of local knowledge and experience, and such groups can arrange unaccompanied visits to sites like working quarries on non-working days because they take safety very seriously, are appropriately equipped, and in addition are fully insured against any mishaps (Fig. 4.2).

Mishaps vary in nature from trivial scratches upwards, but a certain amount of common sense (or in many cases following the rules!) avoids more serious outcomes. In quarries, keeping away from towering faces bristling with shattered overhangs is a good way of staying alive. The author has done a certain amount of rock-engineering over the years, and one such job involved scaling loose rock from a steep 40 m rock-face in a disused dolerite quarry. The face featured a double overhang about halfway up with two large and suspect-looking blocks jutting out a couple of metres each. Anticipating a strenuous effort levering the blocks off with a long steel wrecking-bar, the tools were lowered on a rope and the

Figure 4.2 Collecting
Members of the Russell Society on an officially-arranged collecting trip in a working quarry. Note the prominent safety clothing worn by all participants.

author followed by abseil, stopping and tying-off just above the double overhang. Leaning out over the overhang, he reached down to the lower block and gave it the slightest of taps with the tip of the bar. It immediately dropped to the foot of the face with a resounding crash! Obviously, to anyone caught underneath, wearing a hard hat would have been an irrelevance (Fig. 4.3).

In working quarries, faces are separated from roadways, etc. with **bunds** made from spoil, their purpose being to trap fallen blocks and to act as a barrier over which it is not permitted to cross. Old quarries – and sea and mountain cliffs – do not feature such degrees of protection, and safety depends on good judgment, looking out for suspect areas and fresh scars indicating that rock has fallen recently and that more may follow.

In working mines, the cavernous **stopes** where the ore is extracted are, these days, often mucked-out by remote-controlled equipment. There is a good reason for this, in that rock stability is again an issue: large stopes in old mines are dangerous places, witness the boulders often littering their floors. A hazard particular to underground mines is **false floors**, which are found in stopes where the floor has been removed and replaced by timbers over which spoil is spread before laying the tracks that the wagons run along. Given that many such floors are over a century old, the dangers should be obvious, but false floors are not always apparent.

Other underground hazards can include deep flooded workings and, in unventilated areas, bad air. Underground shafts (known as **winzes**) may

Figure 4.3 Dangers in disused quarries

Beware of rock-faces when collecting! The image shows the face of a disused dolerite quarry; note the prominent overhang near the top of the photo, comprising two huge blocks with square faces on their undersides. The face is in fact foreshortened and this area is about halfway up. The right-hand block fell away following a single, very slight tap with an iron bar!

be equipped with ladder and platform systems, but how old are they? Stacked piles of waste rock (known as **deads**), supported by large timbers, are a common sight in old mines, but how good are those timbers? These are questions that only experience has much chance of answering, and people can still get caught out. Like rock-climbing (or driving, for that matter), experience helps to lower the risk, but the risks can never be eliminated completely. Teaming up with someone experienced in exploring old mines underground is a good way to develop competence in assessing hazards.

At surface, the most serious hazard that most old mines present is open shafts. Mostly, these are clearly fenced off, but, especially with very old mines, there may be unrecorded shafts, or shafts that have partially collapsed and have become hidden as a consequence. In the 1980s the author, while pursuing his postgraduate research, was studying a particular mine and the abundant mineralization on its tips. One morning he drove up, only to find a shaft had opened up at the precise, flattened spot where he always parked his car: it was a large vertical shaft that would have easily swallowed a minibus let alone a car, going some 25 m down to deep water. Other shafts may be partially disguised by bushes and other undergrowth, which is a good reason for not letting your dogs or small children run around at old mines (Fig. 4.4).

Mine tips have their own hazards, such as large boulders that can tip over onto the unwary. But all collecting sites have one thing in common: the dangers associated with hammering. Any rock-type can produce splinters when

Figure 4.4 Mineshaft
Hazards at old mines: the location of this shaft was not precisely known, until one night in the mid-1980s it opened up; prior to that, the spot was a forestry track turning-circle where people (including the author) parked their cars! The photo was taken about a decade later, hence the vegetation.

hammered but hard, brittle boulders of silicified rock, quartz or rhyolite are especially nasty in this respect. Safety-goggles protect the eyes against such things, and strong leather gloves guard the hands, but the best bet is to try to avoid causing splinters to hit yourself or other people. The risk can be minimized (but not eliminated) by swinging the hammer away from yourself and making sure nobody is situated in that direction; in general, if you are going to take a sledgehammer to a boulder,

firstly make sure there is nobody anywhere near to you.

When undertaking geological fieldwork along rocky shorelines, weather and tides are highly relevant. Purchase a set of tide-tables for the area in question, time your visit carefully, and follow the tide down as it goes out, noting any cut-off points and allowing adequate time for your return before the incoming tide again bars progress. Large swells and storms are dangerous and are to be avoided – luckily the internet provides wind and swell forecasts via official forecasting agencies and on websites dedicated to surfing and other watersports, so there is no excuse for getting caught out. Likewise, when collecting in mountainous country, consult the relevant forecasts for hill-goers. Poor weather in the hills, when visibility is severely reduced and navigation is impossible without map and compass, is in any case going to make it very difficult to identify outcrops of interest.

4.3 Mineral collecting ethics

So, assuming you have been given permission to collect at a site, have been briefed properly about safety, have the appropriate kit and have picked a decent day to visit, it is time to consider ethics. This is a thorny area! It can lead to furious debate and polarization; however, it may be boiled down to a few simple principles that, if followed, will hopefully leave the individual free from criticism.

First and foremost, it has to be accepted that mineralogy is a specimen-based science, and the more academic the nature of the work, the more that point holds. Time and again, research has shown that mineral assemblages tend to be far more complex and diverse than initial examination of hand specimens will reveal. This complexity and diversity can have important implications for the proposed genesis of a mineral deposit, the production of the magma that cooled to form an igneous rock, or the degree of regional metamorphism of an area – and many other examples exist. Determining such things requires samples; there is no alternative. Samples may be examined through the binocular or scanning electron microscope; thin and polished sections may be prepared from them; geochemical analyses and x-ray diffraction studies may be carried out on them.

That said, it should also be considered that in all cases, we are talking about finite resources. In some instances – a body of granite that is several miles across, for example – it doesn't matter. In others, such as an isolated mineralized outcrop that uniquely displays useful geological information, it very much matters. Careless sampling can remove or obscure that information, severely degrading the resource. In the UK, many such sites have a degree of legal protection as Sites of Special Scientific Interest (SSSIs), whereby such defacement could lead to prosecution for the offending party.

Much is made in some circles of what is referred to as 'over-collecting', meaning the removal of large quantities of mineralization from a site, with the term implying a degree of greed. However, it's not quite that simple. Firstly, there are, scattered around the world (including in the UK), perfectly legitimate specimen-mines, worked just as legally as mines of metals, coal and so on. Their business, by definition, involves the removal of large quantities of material for preparation and sale. On the other hand, there are instances where people have done the same thing in an unlicensed manner, which can lead to bad feelings, especially with respect to landowners.

Mineralogists who study microscopically-crystalline minerals (microminerals) are working with an entirely different medium. A micromineral collector, be they academic or non-academic, needs to locate material of interest and collect a quantity of it. It will then need to be taken home or to the lab, superficially cleaned and broken up carefully. The minerals of interest will tend to occur in scattered cavities, often as very delicate crystals. Specimens, which tend to be small, will then be sealed away in plastic boxes, to avoid the damage that dust-contamination would do. It is easy to understand how difficult these procedures would be on a windy day in the field, with the rain lashing down!

If collecting from a site that produces well-crystallized specimens, do think about leaving some prospective material behind so that others can have the same enjoyable experience. This applies to all natural sites and some artificial ones such as road-cuttings. However, and paradoxically, there are occasions where excellent material has been found, and where the collector realizes that the rest of it will be destroyed. This applies particularly in working quarries: if it can be established that a pile of newly-blasted boulders will end up being crushed in a matter of days, then feel free to collect as much as you can (this is known as '**rescue-collecting**'). But be prepared to share the material with others and with your local museum – the nearest one with a geology department, that is. The material may not just be pretty: it may be of significant research importance and is of no use to anybody if left sat in trays, wrapped in newspaper, in a shed for decades.

Mine-tips are an important source of mineralogical study-material, although the material that is often of most academic interest – the lumps of primary ore – tend not to be the most sought-after specimens. Non-academic collectors often seek the more colourful and crystalline secondary minerals, which can involve digging through a tip. This can annoy some people, but it needs to be pointed out that material mined underground and tipped at surface will gradually degenerate due to weathering until it is useless. It could therefore be argued that collectors working through such tips are rescue-collecting in the long term. It all depends on what happens to the material collected afterwards.

Few other branches of science are so locality-specific as mineralogy, and it is of foremost importance to record the locality of a specimen immediately. Unlocated specimens are of no research value and, if you are money-minded, are worth a fraction of their located equivalent. In some cases, the mineral assemblage present on a specimen is so characteristic of a site that its provenance may be safely deduced by an experienced mineralogist, but such cases are not common at 100% confidence levels. So label your samples with locality (including a grid reference) first and foremost – identification comes second, because it can be done at leisure.

4.4 Care, cleaning and curation of collected material

Protect your samples from damage (if crystallized) and from cross-contamination (if collecting from more than one locality). For rock- and ore-samples, use strong, A4 or bigger heavy-grade polythene bags that can be stapled shut and numbered in thick permanent marker-pen against an entry in your notebook. Mineral

specimens often need more tender loving care than this; take kitchen-towel and newspaper as standard wrapping and bubble-wrap if expecting to collect particularly fragile material.

When you get your finds home or to the lab, the next stage begins: preparation. In all cases this first involves cleaning. Rock and ore-samples simply need a good scrub under the cold tap with a stiff brush prior to being sectioned or otherwise prepared for analysis. Crystallized minerals require a different approach, because inappropriate techniques can completely wreck a specimen at this stage. This may be because the crystals are very delicate or because they are water-soluble. The matrix may have cracks running through it. Very delicate crystals are best left uncleaned; careful field-collecting is the key here. Less delicate material may be carefully rinsed under a gentle stream of cold water; areas of dirt can be gently removed using soap and a soft paintbrush. Often, specimens collected underground have a coating of fine clay. One trick in this case is to leave the specimen somewhere dry until the clay has dried out completely. A soak in cold water will then see the clay flake-up and drop off.

Water-soluble minerals such as chalcanthite (hydrated copper sulphate, $CuSO_4.5H_2O$) or halite won't do at all well under the tap, for obvious reasons. A cotton-wool bud soaked in alcohol is one possibility, but careful field-collection is the imperative thing here. Specimens at risk of falling apart due to cracks need to be stabilized with a strong but runny glue that will penetrate down into the cracks and hold the whole thing together.

Specialist mineral-preparation labs also use a variety of chemicals to further clean specimens. Dilute **hydrochloric acid** will dissolve unwanted carbonate such as calcite, which may be obscuring other more important crystals. When collected in the field, crystals are not uncommonly iron-stained, and **oxalic acid** is very good at removing such stains. Gold-bearing quartz specimens are sometimes even enhanced by dissolving the quartz with **hydrofluoric acid**. All such reagents – and hydrofluoric acid is the worst of the lot – are potentially dangerous if mishandled due to their toxicity, fumes and corrosiveness acting in various combinations. Such work is best left to experienced laboratory technicians: it may be possible to arrange to get a specimen so cleaned, but it will cost you, so it is probably only worthwhile if you have found a piece of exceptional quality that may be sold at a later date if you need to recover the expenditure (see Fig. 4.5A, B.

Storage requires some thought if your specimens are going to remain in good condition. Some minerals, like quartz, are so chemically unreactive that all you need to do is prevent crystals getting bumped. But others can become a real headache because of their propensity to react chemically with their environment. Some minerals are strongly **hygroscopic** – that is, they attract water molecules present in moist air, absorb them and start to dissolve – halite being a classic example. The only sure-fire way of preserving specimens of such minerals is to keep them in airtight containers with even stronger water-absorbers such as silica-gel or **calcium chloride**. Conversely, some other minerals readily **dehydrate**. When

A

B

Figure 4.5 Quartz – (A) uncleaned and (B) cleaned
Two hand-specimens of quartz from the same South Wales quarry. When found, they tend
to be heavily coated in iron oxides. Keeping them this way maintains their research value:
alternatively, for aesthetic purposes the coatings may be removed by soaking them for several
days in a solution of oxalic acid.

this happens, the surface of a formerly-clear crystal will become covered in a powdery substance. A classic example is the mineral borax, hydrated sodium borate ($Na_2B_4O_7 \cdot 10H_2O$): this readily dehydrates, and in doing so, converts itself into the similar mineral tincalconite ($Na_2B_4O_7 \cdot 5H_2O$). In some cases, the dehydration process leaves a specimen crumbled to fragments. Again, strict atmospheric control is of key importance when conserving such material.

Sulphides are unstable to air and moisture, although the reaction-rate varies from species to species and even from specimen to specimen of the same species. The latter variation is a particular feature of the iron sulphides such as pyrite and marcasite which, because they are so common, present a major headache to curators of both personal and public collections. **Pyrite-decay** is a familiar term to museum curators, and if not noticed in time it can wreak havoc. This is because when pyrite reacts with oxygen and water a by-product is sulphuric acid, which can then start attacking other specimens in the vicinity. The commencement of pyrite-decay can be identified by the appearance of a powdery yellowish **efflorescence** on the surface of a specimen, and by a sulphurous smell. Once it has set in, the only viable course of action is to isolate the specimen from the rest of the collection. Keeping it in a dry environment may help stabilize it – sometimes. The author has observed that in some specimens the process is also prevented, or at least slowed, by keeping the storage temperature as low as possible, although non-academic collectors might find it hard to persuade their families to let them have a shelf in the fridge for their pyrite specimens! The variability of pyrite specimens with respect to decay is an interesting topic, but a general observation of the author is that those formed in higher-temperature geological environments tend to be less prone to it, although this is perhaps an over-generalization (Fig. 4.6A, B.

In all cases, specimens kept out in the open will in time become covered in dust. That's not a problem with robust specimens like big chunks of amethyst that can be jet-washed every now and then. At the other end of the scale, a vug lined by silky 5 mm malachite crystals is going to be ruined by the slightest dust contact. Such specimens are best kept in sealed plastic containers except when they are being examined. For collection storage, metal drawers are better than wood: some kinds of timber (especially oak) give off acetic acid fumes that can corrode susceptible specimens. In well-organized drawers, each specimen will sit in its own cardboard tray. These are readily available and cheap, and they prevent specimens from banging into one another as a drawer is opened or closed. For display, due to the aforementioned dust, a glass case is the best option, and there are sizes and styles to suit all tastes. They also keep unwanted tactile examiners at bay – people who will stroke delicate crystals to see what they feel like, trashing the specimen in the process. There is also a safety element in that some minerals are hazardous.

Hazardous minerals include those that are poisonous; many metalliferous minerals are toxic if ingested, and some extremely so – so give your hands a good scrub after handling them. Others may be dangerous because their

Figure 4.6 Pyrite-decay
Pyrite-decay: in the first image (**A**) it has just commenced in earnest. The specimen was then placed outside in hot and humid weather for just 24 hours and the second image (**B**) shows the result. Humidity is what causes the decay, but heat seems to accelerate the rate of the process.

crystals consist of tiny fibres that are easily inhaled if disturbed. These can cause lung damage, and some minerals, such as those belonging to the asbestos group, are known carcinogens. Storage should involve a sealed plastic box, or better still, avoid collecting such things. Laboratories – where such things are scientifically studied – have sophisticated environmental controls and equipment where the risk of contact is minimized. The same cannot be said of your spare room at home. The same points apply to strongly radioactive minerals (I'm not trying to put anybody off mineralogy here, but beginners need to be aware). There are three key things to avoid with radioactive specimens: dust inhalation, ingestion and breathing in of the decay daughter-product, radon gas. Cleanliness and careful handling/storage are again key in the first two. With respect to radon, a well-ventilated storage place is needed. Museums often keep radioactive specimens in a room where the air is completely changed on a frequent basis – every few seconds. Low-level radioactivity is, by contrast, a lot more common than the average person might think. Many granites are weakly radioactive, for example. It's the really hot rocks like high-grade uranium ores where special care is of importance.

4.5 Identification of collected material

With experience, many minerals may be identifiable simply via the process of recognition: the brain registers colour, crystal system and habit, associated species and so on. But to reach that point the mineralogist must first embark upon a steep learning curve. The properties all minerals share, as explained earlier in this book, need to be understood and applied both in the field and back at home. That way, you can, for example, confidently state that you have found a bluish-grey mineral with a bright metallic lustre that crystallizes in cubes, has a perfect cubic cleavage, has a hardness between 2 and 3 on the Mohs Scale, is very dense and has a leaden-grey streak. A good mineralogial reference-book – or discussion with another, more experienced collector – will quickly lead you to the identification of the mineral – the common ore of lead, galena.

Staying on the theme of minerals containing lead, understanding **associations** is very important. Common secondary lead minerals that form when galena is weathered include cerussite (lead carbonate, $PbCO_3$) and pyromorphite (lead chlorophosphate, $Pb(PO_4)_3Cl$). They often occur together simply because the carbonate and phosphate anions are very common in natural systems involving rainwater, rocks and weathering. Less common, but still widespread, is the lead molybdate, wulfenite ($PbMoO_4$), but its association with cerussite, and especially pyromorphite, is so frequent that finding orange tabular crystals occurring on pyromorphite means that you have almost certainly found a specimen of wulfenite.

Other minerals can be a lot harder to identify. Take, for example, the lead sulphate-carbonate compound, $Pb_4SO_4(CO_3)_2(OH)_2$. This is represented in nature by no fewer than three different polymorphs, comprising leadhillite (monoclinic), susannite (trigonal) and macphersonite (orthorhombic). The latter is incredibly rare, but leadhillite and susannite are not infrequent occurrences in weathered lead ores. Both have a similar colour and with a wide range of habits, they can be impossible to distinguish by visual means alone. This is where professional assistance becomes necessary; locate your nearest museum or university with a geology department and see if they will identify your sample for you. The analytical technique known as **X-ray diffraction (XRD)**, which can determine a mineral's identity via its detailed and unique crystal structure, is typically the method used. A brief explanation of the technique will serve to show why it is so useful.

X-rays, discovered in the late nineteenth century, are a form of radiation with very short wavelengths varying from 0.1 to 10 nanometres – way, way smaller than visible light. Early in the twentieth century, physicists William Lawrence Bragg and his father, William Henry Bragg discovered that when bombarded with X-rays, crystalline solids such as minerals produced unique patterns of reflected X-rays: at certain X-ray wavelengths and incident angles, crystals produced a series of well-defined peaks (Bragg Peaks) of reflected radiation. Bragg junior worked out that this could be explained by modelling a crystal's internal atomic architecture as sets of discrete parallel planes, the planes in each set separated by a constant parameter that he termed 'd', measured in **ångströms** ($1Å = 0.1$ nm). An X-ray that reflects off the surface of a crystalline substance has traveled less distance than one that reflects off a plane of atoms inside the crystal. The penetrating X-ray travels down to the internal layer, reflects, and travels back over the same distance

before being back at the surface. The distance traveled depends on the separation of the layers ('d' – often referred to as the 'd-spacing') and the incident angle ('θ') at which the X-ray entered the material. Bear in mind that X-rays arc waveforms whose wavelengths ('λ') depend on a variable that is controlled by the researcher – the X-ray source. Typically, for mineralogical purposes, the X-ray tube that is used contains copper as the source: in the tube, high-voltage electrons are made to collide with the source, which then emits the X-rays. Those emitted by copper have a wavelength of 0.154 nanometres. The X-ray diffractometer passes these X-rays through a diffraction grating to produce a narrow beam of X-rays that falls onto the surface of the crystal being investigated in the spectrometer. The X-ray source and the detector that picks up the reflected X-rays are both stationed at the same distance from the sample surface.

Now, for these internally reflected X-rays to be in phase with the X-rays that were reflected off the crystal surface, they need to have traveled a whole number – an integer ('n') – of wavelengths while inside the material. Watch the waves crashing into a sea-wall on a rough day. There is the large long-wavelength ground-swell rolling in off the ocean and there are the shorter but sharper waves caused by the gale. Mostly they are slightly out of phase with one another, slopping and splashing about ineffectively. But every now and then both are in phase, and they complement one another, making for a much bigger wave that comes thundering over the sea-wall. This is constructive interference at

work. It is the same with X-rays being reflected by a crystal: when the X-rays, reflecting off plane after plane within a crystal, all separated by a constant distance – the mineral's d-spacing – are in phase with each other, they will constructively interfere and add together, giving a strong signal to be picked up by the detector. Out of phase X-rays, by contrast, will partially to wholly cancel each other out, producing a weak to barely detectable signal.

Bragg senior's experiments made him realize that when changing the angle of the incident X-rays, very strong reflections occurred at specific angles, depending upon the orientation and spacing of the different planes of atoms that make up the crystals of any specific mineral. All of this gets us to the following:

$$n\lambda = 2d\sin\theta$$

That is Braggs' Law, a major part of a groundbreaking era of materials search for which the Braggs were awarded the Nobel Prize for Physics in 1915. The equation shows that if λ – the wavelength of your X-ray source, and θ – the angle of incidence – are known, which they are, then the technique will give you the unknown – d – thereby identifying your mineral (Fig. 4.7):

$$2d = n\lambda/\sin\theta$$

These days, the process is automated to a large extent: the sample, be it a single, painstakingly-mounted crystal or a ground-up rock or mineral fragment, is mounted in the spectrometer and either it is rotated through a range of θ values or the X-ray source is rotated similarly relative to the specimen. The data output is typically plotted on a graph whose x-axis is the angle at which the signals are detected (2 θ, in degrees) and the y-axis is reflective intensity, i.e. signal strength, in counts per second or percent. A typical graph consists of a series of peaks of

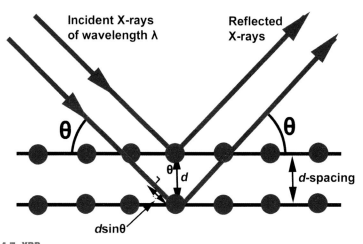

Figure 4.7 XRD
Diagram showing how Bragg's Law works. The wavelength of the incident X-rays is known (λ) as is their angle of incidence (θ). Because the reflected waves are only in phase when 'n' is an integer (1, 2 etc), the unknown, 'd', may be determined.

varying intensity against a relatively flat background; the intensity and relative positions of the peaks form patterns unique to each and every mineral. Back in the author's postgraduate research days, the graph would churn out in rolls of continuous paper from the spectrometer – provided it had not run out of ink. Today, the data appears on a monitor and specialized software examines it and matches it to known mineral species, all of which it has on a database.

So, back to our impossible-to-tell-apart leadhillite and susannite. X-ray diffraction of the sample will give one of the results shown in Table 4.1: The data are sorted so that the most intense peak is at the top and the least intense peak at the bottom. As can be seen, there are subtle differences in the d-spacings (and thereby 2θ values) for the major peaks, and more obvious differences in the size and position of the smaller peaks. Note that the table only includes the ten most intense peaks for either mineral: a typical XRD dataset may contain dozens of minor peaks.

From time to time, mineralogists will turn up a crystalline mineral whose XRD trace fails to match any known species. Quite literally, they have turned up a 'known unknown' – a new mineral! However, a lot of complex analytical chemistry and crystallographic work then lies ahead in order to describe the mineral and give it an official name. It can take years between recognizing that a mineral is unknown to science and publishing a full description of it in the peer-reviewed scientific literature. One major difficulty is that many such finds consist of microcrystals a fraction of a millimetre in size; the practical difficulties of extracting pure samples of such material from matrix and working on them are not to be underestimated.

Help with mineral identification is easier to obtain today than ever before, with a plethora of high-quality and well-illustrated books and journals. In the UK, there are currently two main journals based around topographic mineralogy (mineral occurrences and the mineralogy of specific localities): these are the UK Journal of Mines and Minerals and the Journal of the Russell Society. Online, there are several mineralogical reference websites, a leading one being Mindat – www.mindat.org – which has a very active online community of non-academic and professional mineralogists, where one can get help and suggestions in mineral identification, including advice regarding where to go to obtain analytical help. Plus, of course, there are the sites where mineral specimens are offered for sale, complete with hundreds of mouth-watering photos of some incredibly spectacular wonders of the natural world, which brings us neatly to the next section.

4.6 Buying minerals
For all collectables there exist dealers, and mineral specimens are no exception. Mineral dealing exists on a number of levels. Firstly there are people whose day-job takes them underground or into quarries and who have cottoned on to the idea that there might be some value in crystallized minerals. Such people were once commonplace in metalliferous and industrial mineral-mining

Table 4.1 Comparing the x-ray diffraction patterns of two polymorphs of the same lead compound – the minerals leadhillite and susannite

Leadhillite			Susannite		
D-SPACING (Å)	2θ (degrees)	INTENSITY %	D-SPACING (Å)	2θ (degrees)	INTENSITY %
3.5529	25.06	65.97	3.5695	24.95	67.41
2.6057	34.42	36.7	2.6186	34.24	35.64
3.5561	25.04	35.78	3.5695	24.95	32.59
3.5532	25.06	34.03	2.9384	30.42	30.54
2.9267	30.54	27.87	4.5355	19.57	20.71
4.5095	19.69	20.6	2.1114	42.83	11.55
3.5555	25.05	17.27	11.572	7.64	9.35
11.5400	7.66	16.23	2.3144	38.91	8.67
2.6025	34.46	16.01	2.0615	43.92	8.38
2.9278	30.53	15.96	2.1114	42.83	8.31
2.308	39.03	13.83	2.893	30.91	7.99

centres, such as Weardale in the North Pennines. However, as the mines have long since closed down, the days of casually popping into local pubs and asking, 'anyone got any spar?' have gone. At one time it was a good way to acquire decent material directly and at fair prices, although it did not exactly meet with full approval from the companies the guys worked for! (Fig. 4.8.)

Commercial mineral dealers fall into two broad categories, although there is often a degree of overlap. Firstly, and most conspicuously, there are the touristy and new-age shops to be found in most towns. They sell, among a wide range of other things, low- to medium- grade affordable material like amethyst or pyrite that is obtained via **bulk dealers** – wholesalers who import it by the ton. They do well because many non-mineralogists still appreciate the beautiful colours of minerals such as malachite or lapis lazuli, and the lustre and form of natural crystals. Such material, which also includes artificially-treated minerals like agate slices dyed blue and 'smoky' quartz that is in fact low-grade rock crystal that has been irradiated, is unlikely to appeal to any serious mineralogist; it is a different branch of the rock-trade altogether.

Secondly, there are the professional mineral dealers supplying the global mineral collecting fraternity. They vary from people who specialize in good representative specimens of rare minerals for **systematic collectors** (e.g. someone who aims in due course to acquire a specimen of every carbonate or arsenide mineral species in existence) through people who specialize in regional mineralogy – for example, minerals from the south-west of England – to people who specialize in high-quality display specimens. Quite a few dealers fall under all three varieties depending upon what they have

Figure 4.8 Fluorite from Weardale
Years ago, when many mines were still operating, specimens like this 20 cm long bed of fluorite crystals could be obtained for cash simply by asking around in the pubs in Weardale in Northern England.

in stock at any given time. They obtain their material via networks of contacts around the world, by visiting mineral shows and by buying up old collections.

Mineral dealing has done an awful lot to preserve important specimens over the past three centuries, and as such it has done a great service to the science of mineralogy; many important museum pieces the world over owe their continuing existence to the fact that they have been carefully passed from collection to collection, perhaps to eventually be purchased from a dealer or be bequeathed to the nation upon the death of a collector. It is perhaps surprising that the trade is frowned upon in some circles, but it nevertheless is. The problematic light in which

mineral dealers are sometimes bathed is, however, similar to that in which collectors are seen by some non-mineralogists, which is dealt with in section 4.3.

That said, in every trade under the sun there exists a rogue or two. Faked mineral specimens, the handiwork of a tiny minority of dealers around the world, has a long pedigree going back two centuries or more. An old nineteenth century classic involved faking large cassiterite pseudomorphs after euhedral orthoclase feldspar from a mine on the north coast of Cornwall. These were in great demand at the time, so that when the supply was getting exhausted, someone had the bright idea of carefully fashioning them from sheet lead, coating them with glue and

then covering them with powdered cassiterite; the appearance was quite similar, as was the density. Today, variations on the theme are still alive and well: loose crystals cunningly glued onto matrix, botryoidal aragonite stained pale blue and passed off as rarer, more desirable species, crystals grown onto matrix in the laboratory, and so on. Experienced collectors will spot such things a mile off, however, and there are even websites (such as www.the-vug.com/vug/vugfakes.html) that blow the whistle on such activities! (Fig. 4.9.)

Serious dealers with a good reputation will not for one second entertain such ideas – they know that their reputation has taken much hard work to build up, and that once stained in such

Figure 4.9 Faked cassiterite
A cunningly-made artificial specimen: a faked Cornish pseudomorph of cassiterite after orthoclase, made out of sheet-lead coated with glued-on powdered cassiterite. Oxidation of the lead has occurred, allowing some of the coating to drop off and thereby revealing the fraud. Copyright: the Trustees of the National History Museum, London.

a way it would be impossible to recover. That's why they are the best people to go to if you are serious about buying a few pieces to add to the collection. But how do you go about buying?

Mineral specimens are more like antiques than commodities such as gold bullion, which have a fixed price – so much per ounce – that only varies according to the whims of the global metals market. There is no hard-and-fast rule regarding mineral specimen prices except, perhaps, the rule of thumb that a specimen is only worth what someone is prepared to pay for it. That is so abstract that it is not particularly helpful to a collector just getting into buying. The best advice is to shop around; most dealers now have lavishly illustrated websites and there is no substitute for careful homework in this respect. Even if you are going to buy by visiting dealers in person, you should be able to develop a good idea of current price-ranges and recognize any extortionately-priced specimens before handing over your cash.

People are sometimes astounded at the prices for which some mineral specimens are sold (Fig. 4.10). However, there are collectors out there who, if they want a 'killer' world-class specimen badly enough, will pay half a million dollars for it. Such acquisitions represent major investments that are expected to at least hold their value over time and likely increase in worth; at the same time, some very respectable material can be acquired for $500 or less (the US Dollar being the general unit of currency in the global mineral trade these days). One good tip is to take advantage when a major find appears on the market, so that there are

Figure 4.10 A high-end mineral specimen
Mineral specimens are available to suit all budgets: this magnificent, 21 cm high cluster of stibnite crystals, from Jiangxi Province, China, is yours for just $23,500!

lots of specimens to choose from. This may represent the results of six months underground somewhere, working a series of cavities lined with crystals. At first, there tends to be a trickle of specimens; later this becomes a flood, and then is the time to buy. In due course the source is exhausted, the supply dries up and the price per piece of the remaining material goes up again. It's a basic case of supply and demand. Good Weardale fluorite specimens now fetch crazy prices, compared to what one could have obtained, having picked through dozens to make a choice, from a miner in any one of the valley's pubs in the 1980s with a £50 budget. That is just one example out of many such situations.

Buying minerals opens up mineral collecting in a number of ways. It allows collectors who cannot get out into the field, perhaps due to physical disability, to continue collecting. It allows collectors interested in an old locality, long since obliterated, to obtain and study material from there. It also makes it possible to build up suites of material from inaccessible sites halfway around the world. Done with care, it can strongly enhance a personal collection.

A final means of acquiring hard-to-get material is by exchange, either with other collectors or with dealers. This is less of an option for complete beginners, but as a collection is built up through time, duplicates are bound to be present. The ideal opportunity to create an exchange-stash will be when a lucky collector is at the right place at the right time, perhaps arriving at a quarry after a recent blast to find boulders in the muckpile full of cavities of well-crystallized rare minerals. Knowing that

it's all going to end up going through the crushers within 24 hours, the collector removes the best specimens – there may be dozens – and after selecting those they want to keep, offers the rest for exchange. Thus a rare find that would have been completely destroyed gets circulated throughout the mineralogical world, which can only be a good thing. The main piece of advice with respect to exchanging is to only offer pieces of a quality that you would expect to receive. Pieces with broken and bruised crystals are not suitable as exchange specimens. However, they may still have value to science, if they are something particularly new and/or unusual. Such material is suited to destructive analysis – it does not need to have any aesthetic quality to yield up good data when acid-digested and analysed, or sawn up and made into thin or polished sections. So, with respect to the ugly specimens left over from your big find, consider your local Museum and University research departments as a repository!

4.7 Museums and minerals

The world's museums hold some of the most important mineral collections in existence, and the UK is no exception. Older institutions, in particular, have collections that were assembled during the heyday of British metal mining, when literally hundreds of mines were at work. The Royal Cornwall Museum in Truro, Cornwall, founded in 1818, is a case in point: it houses most of the Rashleigh Collection. Composed largely of minerals from the south-west of England, and especially Cornwall, it is almost certainly the most complete contemporary collection from that area. It was collected by Philip Rashleigh

(1729–1811), who lived near Fowey in southern Cornwall, which he represented as an MP in Parliament for nearly forty years. During his mineralogical career he obtained over 3000 mineral specimens, many of which are world-class in their quality or crystal size and are meticulously catalogued. He produced two volumes illustrating the collection, entitled *Specimens of British minerals, selected from the cabinet of P. Rashleigh, of Menabilly, in the County of Cornwall … with general descriptions of each article,* which were published in 1797 and 1802. Following his death, the collection remained with the family in Cornwall and in time ended up at the museum (some specimens are also in the Natural History Museum, London). It is quite rare, indeed, for a collection of this antiquity to survive largely intact for such a long time.

Other important collections include more recent but extensive suites of material such as the King Collection, assembled by Bob King (1923–2013), held at the National Museum of Wales, Cardiff and the Russell Collection, the work of Sir Arthur Russell (1878–1964), preserved at the British Museum (Natural History). The latter institution also holds the collection of Arthur W. G. Kingsbury (1906–1968), a contemporary of Russell and a very active collector in several districts of the UK. Kingsbury published a large number of papers in the mineralogical literature, describing first British occurrences (some 50 in all) of a number of rare minerals, especially during the 1950s. However, in recent years staff at the Natural History Museum have unearthed something quite remarkable, in that a number of these reported occurrences were no such thing; the specimens described matched perfectly with material from known overseas localities. This dubious practice, which is thought to have begun around 1951, has cast doubt on the provenance and scientific value of the entire Kingsbury Collection, requiring a meticulous specimen-by-specimen examination to verify pieces. In turn, in-depth reviews of the mineralogy of various sites have been necessary; for example, from just one such review:

The claimed occurrence of adamite at Sandbeds mine represented the first British report of the mineral when L. J. Spencer noted it on the basis of a specimen 'collected' in 1951 by Arthur Kingsbury (Spencer, 1958; Cooper and Stanley, 1990). This specimen and others supposedly collected by Kingsbury at Driggith mine were discredited by Ryback et al. *(2001, p.52), who proved beyond reasonable doubt that they came from Lavrion in Greece.*

As can be seen, such activities, once committed, permeate out into the mineralogical literature, and continue to do so until the truth comes to light. Fortunately, such widescale and systematic fraud is extremely rare – to the extent that in the cited case, a whole raft of papers have been published clarifying aspects of it. But why do it in the first place? The fraud was certainly not based around money; the collection was donated. Perhaps, it has been surmised, he was attempting to artificially inflate his status as a successful collector against his contemporaries; but the true reason will perhaps never be known. Today, with a much greater number of active and knowledgeable collectors and the thousands of images of minerals from worldwide localities that can be studied online, such a thing would be almost impossible to perpetrate.

Back to the vast majority of genuinely collected material, though: and if you develop an important collection of mineral specimens through your lifetime, it is well worth considering willing their donation to a museum on your demise. This is particularly the case if your surviving family have no interest in the minerals; scientifically-important material can then run the risk of being dumped as bric-a-brac. A friend of a friend who works in one of the UK's major museums once showed him a specimen of gold from a Welsh mine (the mineral assemblage was absolutely diagnostic) that was on a par with any known examples. He had found it among 'some rocks' in a Welsh charity-shop and bought it for a couple of pounds. That one specimen, it turned out, was worth the rest of the shop's contents put together!

Museums may also have the ability to purchase collections, and if you are thinking of selling up, and you have an important suite of minerals from a particular area, it is well worth getting in touch; if they go to a mineral dealer it is likely that the specimens will be scattered to the four corners of the planet, where they may look great in people's cases but have lost much of their value to science. This is because it is the overall mineralogy – the whole assemblage – that is important when studying the formation of a specific mineral deposit at a specific locality.

Examining the mineral collections of museums tends to work in two ways, because specimens are divided into those on display in public galleries and those stored behind the scenes. Walking around a major mineral-gallery is often a jaw-dropping experience, with cases full of neatly arranged and labeled well-crystallized specimens; over the decades such first encounters have inspired many young minds, leading in time to careers in mineralogy or serious collecting activities. To the established mineralogist, the area of the museum not normally open to the public, where all the curatorial, research and conservation work goes on, is equally important. Here, by prior arrangement, material can be examined – for example, using a binocular microscope – for research purposes. This may involve looking at specimens collected from a certain locality and comparing them to what you have found there. It may involve research into a suite of samples collected years ago at a locality that has long since disappeared through the ravages of time and/or Man. Or the collector may be bringing in a sample suite they have collected for help with identification, as noted in the identification section earlier. Our museums are an invaluable part of our society, and it is important to support them, both on a personal level and at a national one.

5 Studying mineral assemblages and parageneses

5.1 Introduction

Much field mineralogy, especially advanced research or mineral exploration, involves the identification and description of mineral associations and parageneses. Certain minerals will tend to occur together, as outlined in more detail in chapters 2 and 3: for example, igneous and metamorphic rocks are identified by the key mineral assemblages that they contain. In coarse-grained rocks and ores, such determinations may be carried out in the field using a hand-lens or even the naked eye. Finer-grained igneous and metamorphic rocks and ores are rather more demanding, requiring examination under high magnification to deduce the minerals present (the assemblage) and the order in which they formed (the **paragenetic sequence**). **Paragenesis** is important because the sequence in which the minerals came together to form a rock or ore can tell us about the changing conditions during that process. In economic geology, there are often cases where a mineralized district carries several generations of vein mineralization. There are numerous cases in which each generation of mineralization looks broadly similar to the others in hand specimen, but only one or two may carry economically interesting levels of commodities such as gold. This is where paragenetic studies can become critically important (Figs 5.1 and 5.2).

5.2 Petrography

Whilst hand specimens can be further examined in the laboratory with a binocular microscope (and many non-academic mineral collectors possess such equipment), in most cases the true nature of the paragenetic relationships between closely intergrown minerals making up an assemblage can only be deduced by examining samples under high magnification, which involves the use of an optical (petrographic) microscope, and in some cases a scanning electron microscope. The geological sub-discipline by which such samples are investigated and described is known as **petrography**. It is a branch of the broader science of **petrology** (which covers more generic investigations into rocks), and it focuses on detailed descriptions of rocks and mineral assemblages, textures, intergrowths, structure and so on. For such examinations, specimens have to be prepared to precise standards. In the majority of cases, this means making sections from the collected samples.

There are three main ways in which sections are prepared, depending on the material under investigation: the resultant prepared specimens are termed thin, polished and **polished thin sections**. Standard thin sections involve samples being cut and then carefully ground, affixed to a glass microscope slide, to a standard thirty micron (0.03 mm) thickness and examining them using a specialist petrographic (sometimes called polarizing) microscope. At such a thickness, the vast majority of the rock-forming minerals are translucent in transmitted light. To study opaque minerals such as most sulphides, native metals and some oxides (i.e. pretty much all ore-samples), a different technique is employed whereby the sample is embedded in hard-setting resin, ground flat on a lap and then carefully polished in several stages (and meticulously cleaned between stages) to an optical-grade finish using **diamond pastes** of progressively finer grade. The resultant polished section can then be viewed using a second type of microscope lighting – reflected light. Transmitted and reflected light microscopes were at one time manufactured as separate pieces of equipment, but modern petrographical microscopes now come equipped to work in both transmitted and reflected light modes. Furthermore, there has been an increased tendency in recent decades for laboratories to produce polished thin sections that can be examined in

Figure 5.1 A straightforward paragenesis
Large (30 cm wide) specimen of coarse-grained vein mineralization from North Wales. Sphalerite and calcite occur in repeated bands from the vein wall (bottom) to its central cavity (top); some of the bands also contain marcasite and galena. The order in which the minerals were deposited can plainly be seen.

Figure 5.2 A more complex paragenesis
From the same area of North Wales as Fig. 5.1, this hand-specimen of quartz and intergrown sulphides belongs to a completely different generation of veining. Chalcopyrite, pyrite, sphalerite and galena are all visible to the naked eye but their relationships to one another are not discernible because they are so finely intergrown. Together, these specimens demonstrate two important things: firstly that mineralized districts commonly contain more than one generation of mineralization, and secondly that sectioning and examination of mineral assemblages using a petrological microscope is essential if the mineralogist is to get the full picture.

both transmitted and reflected light modes. These are more painstaking and therefore expensive to prepare, but for the petrographer undertaking the research they make things a lot more efficient.

5.3 Petrographical microscopy – transmitted and reflected light

Petrographic microscopes vary in their complexity from basic teaching microscopes up to the highly expensive tools used for advanced research, and which are typically equipped with high-quality camera equipment. However, all share the same basic features that first-year undergraduates will rapidly become familiar with: **eyepieces** down which you look, **objective lenses** (often three or four, from 10× up to 100×, mounted into a rotatable head) and a **stage** upon which to place the sample. Between either of the light sources (beneath the stage for transmitted light, above the stage for reflected light) and the stage, there is a **polarizer**; normally, light-waves vibrate outwards at right-angles to their overall path (think of the waves that develop on a fairly taut rope when it is shaken). The polarizer blocks all but those vibrating in one orientation (typically east–west); this is known as **plane-polarized** light. Petrographical microscopes are also equipped with a second inline polarizing filter, known as the **analyser**, situated between the eyepieces and the objective lenses, which may be rotated. At right angles (i.e. north–south) to the first polarizer, in a situation termed 'crossed polars', all light is blocked – until a section is placed on the stage. Why is this filter so important? Because many minerals polarize light as a result of their structural chemistry. Often, the light is split and polarized into two different planes of vibration – the double refraction, or **birefringence**, of Iceland Spar calcite being the classic example. Under the microscope, what happens when a thin section containing a birefringent mineral is viewed under crossed polars? The polarized E–W light hits the base of the section, but as the light passes up through the mineral grain it is either refracted or double-refracted, so that it is no longer of an E–W orientation, but will be vibrating in two different planes as it emerges from the top surface of the section and travels up through the objective lens. The light then hits the analyser, which blocks all E–W light through but lets some of the newly oriented light through, on up through the eyepieces to the observer's eyes.

The colour of a mineral under plane or cross-polarized light is a function of what light wavelengths the mineral can absorb, across the **visible spectrum** from violet (wavelength 390 nanometres) to red (wavelength 760 nanometres). A mineral that lets all wavelengths between these two extremes pass through will look colourless. A mineral that lets none pass through (i.e. it absorbs them all) will appear opaque. Minerals that absorb specific wavelengths of light will appear to have specific colours under plane-polarized light. Under crossed polars, the situation becomes rather different. This is because two rays of split and refracted light moving through a mineral grain will typically do so at slightly different speeds, so when they move back out into the air above a section they will be out of synchronization. The analyser will combine them into one plane of movement, but one in which there are two wavelengths interfering with one another, producing what are known as **interference-colours**, which are often very different to the 'normal' colour of the mineral (Fig. 5.3A, B).

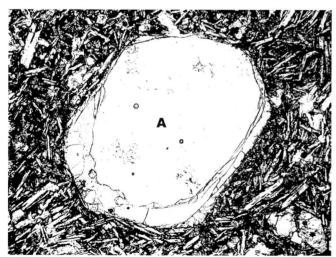
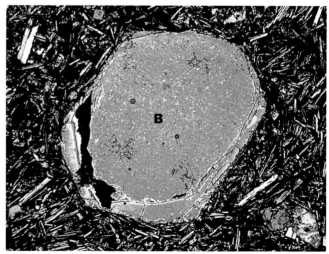

Figure 5.3 Olivine – (A) PPL and (B) XPL
A 15 mm wide phenocryst of olivine in a thin section of basalt viewed in plane polarized light, where it is colourless, whereas under crossed polars it displays bright second-order interference colours.

Minerals that show interference colours in thin (or polished) section are termed **anisotropic**. Minerals that remain black under crossed polars regardless of orientation are termed isotropic. Only minerals crystallizing in the cubic system exhibit isotropy at all orientations. Anisotropic minerals will appear dark at certain orientations as a microscope stage is rotated; this occurs four times during a complete 360° rotation and is known as **extinction**. Extinction angles are diagnostic in mineral identification. Some minerals can show isotropy in sections cut exactly parallel to certain crystallographic planes; this applies to tetragonal and hexagonal minerals in sections cut parallel to the basal crystallographic plane. However, a section cut through an intergrowth of mineral grains, such as will typically be found in a granite, for example, will tend to pass through the various crystals at all angles.

The interference-colours of anisotropic minerals are useful in mineral identification, provided the thin section has been properly prepared to the standard thirty microns thickness. They are classified into a number of Orders as follows:

Table 5.1: Orders of interference-colours of anisotropic minerals

Order	Colours
First	grey, white, yellow, red
Second	violet, blue, green, yellow, orange, red
Third	indigo, green, blue, yellow, red, violet
Fourth	pale pinks and greens

The greater the birefringence of any one anisotropic mineral species, the higher the Order of its interference-colours, although the following caveat applies: since the maximum and minimum refractive indices of any mineral are oriented along precise crystallographic directions, the highest interference-colours will be shown by a section through a mineral grain that has both maximum and minimum refractive indices (i.e. the maximum birefringence) in the plane of the section. At any other orientation, the mineral will show lower birefringence and lower-Order colours. Typically, in, say, a section cut through an igneous rock, the crystals will be present in all orientations so that a range of birefringence values will be displayed by the crystals of any one mineral, from its maximum down to its minimum. It is thus necessary to describe birefringence by the highest colour observed for that mineral: thus a section full of olivine crystals will show a mixture of grains in extinction and grains showing the bright second- and third-order interference-colours that it should display when ground down to 30 microns thickness. One can now appreciate why having that 30 micron section thickness as a standard in petrography is important: a mineral with a certain birefringence value displays different interference-colours at different thicknesses. Look at a badly-made thin section that has been ground down to 15 microns and those olivine crystals orientated so that their maximum birefringence is displayed will now unhelpfully be showing anomalous first-order reds.

In polished sections viewed under reflected light through crossed polars, cubic minerals again appear isotropic, but in non-cubic species, **anisotropy** is a distinctive and diagnostic property that is described in terms of its strength, from weak (difficult to detect) through to strong (in which pronounced changes in brightness and possibly colour occur as the stage is rotated).

Minerals may also be **pleochroic** (both in transmitted and reflected light) transitioning from one colour to another as the stage is rotated. Biotite is a good example: depending on its orientation relative to the plane-polarized light, it absorbs more or less of that light, so it will turn from brown to yellow and back as the microscope stage is rotated through 180°. In reflected light, another property, **reflectance**, is important and diagnostic. Reflectance (or **reflectivity**) is the amount of light reflected from a mineral's polished surface and is expressed as a percentage value. Absolute values are hard to estimate by the eye, because a mineral of a certain reflectance can look brighter or duller depending on its surroundings. For example, sphalerite (zinc sulphide, ZnS) has a reflectance of just 15%; it appears a dull grey in polished section. Intergrown galena (lead sulphide, PbS) with a reflectance of 43% looks bright white against sphalerite. However, seen against sulpharsenides or sulphantimonides, such as ullmannite (NiSbS), which tend to have much higher (>50%) reflectance values, galena appears a dull grey (see Figs 5.4 and 5.5A, B). These relative reflectances can deceive the inexperienced eye. Some anisotropic minerals also exhibit changes in reflectance upon the stage being rotated; this is known as **bireflectance** and is described as weak to strong. Non-opaque minerals in polished section may display distinctive **internal reflections**, especially under crossed polars.

Figure 5.4 (left) Pleochroism
Biotite flakes (shades of brown) in a thin section of granite (field of view 6 mm): the crystals exhibit pleochroism with colours varying with individual crystal orientation.

Figure 5.5 (below) (A) Galena-sphalerite and (B) Ullmannite-galena
Galena in polished section has quite a high reflectance at 43%, which makes it look bright white against the much less reflective (15%) sphalerite. However, against minerals with higher reflectivity (in this case ullmannite, NiSbS), the galena looks a rather dull mid-grey, the point being that the amount of reflectance of a mineral that is perceived by the human eye depends critically on that mineral's surroundings.

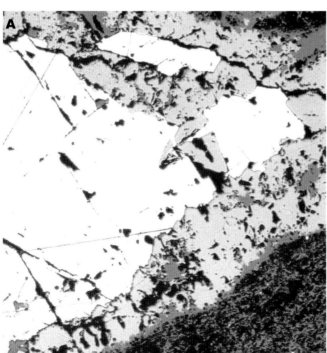

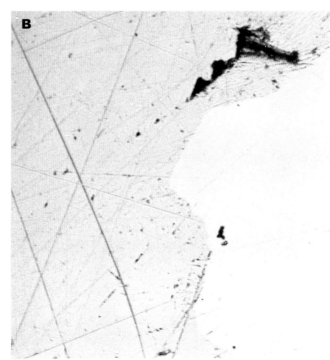

In thin sections, some minerals really stand out from the background. This property is known as '**relief**' and occurs as a consequence of the mineral having refractive indices that have a marked difference with that of the special resin (refractive index of 1.54) used to mount the rock to the slide and the cover-slip to the rock. Relief in polished sections is in contrast a topographical property; consistently producing low-relief polished sections is a measure of the skill of the section-maker, for relief in polished sections is caused by softer minerals being more easily abraded than harder ones. Polished sections with high relief are useless, for

relationships along grain-boundaries between hard and soft minerals – for example, gold (soft) veinlets in pyrite (hard) – become obscured. In really bad cases, small grains of soft minerals may be impossible to observe (Fig. 5.6A, B).

Artefacts produced in polishing may in other cases be useful and diagnostic. A good example is galena; soft, brittle and with a perfect cubic cleavage, the mineral readily and unavoidably 'plucks' during the polishing process, leaving rows of distinctive triangular pits. Thus, galena may be identified under the microscope at a glance. In cases where the sample material is friable – perhaps because it is weathered – thin and

polished sections can only be prepared once the material has been vacuum-impregnated with special epoxy resins.

5.4 Scanning electron microscopy

The next step down the magnification road is the **scanning electron micro-scope** (usually referred to as the SEM), which images samples using a focused beam of high-energy electrons. The interaction of the electrons with the sample yields information about its morphology, composition and structure. Whilst petrographical microscopes can be found lined up on lab benches, SEMs occupy entire rooms, and sessions with them are typically booked in advance

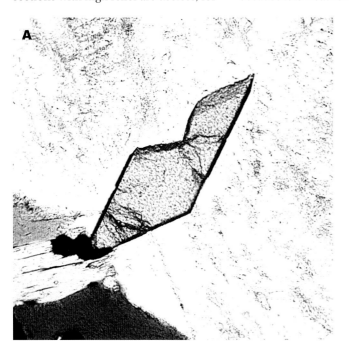
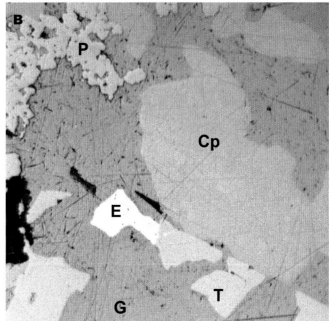

Figure 5.6 Relief in thin (A) and polished (B) sections
Relief means different things in thin and polished sections. In thin sections, the term 'high relief' means that the mineral has a much greater refractive index that either the mounting resin (1.54) or the groundmass, so that the mineral grain or crystal really stands out, such as this central crystal of titanite in a thin section of granite (field of view 1 mm). In polished sections, a mineral with high relief is, because it is relatively hard, standing proud of the section surface. In the illustrated example, pyrite (P) has high relief against the much softer galena (G). The three other minerals present, chalcopyrite (Cp), electrum (E) and tuckeite (T) are closer in hardness to the galena so that their grain-boundaries have polished better.

and supervised by a technician, to avoid anything going wrong. This is unsurprising when one considers that they can cost hundreds of thousands of pounds.

How does a SEM work? Firstly, the sample (either a chip of rock or a polished section) has to be rendered conductive over its surface and grounded. The reason for this is to prevent electrostatic charges building up and confusing matters: the procedure involves coating the sample in a thin layer of carbon, gold or other conductive material. The sample is then placed rigidly in a specimen holder within the specimen chamber, a container that may be pumped down to vacuum, which is where the business end of the apparatus is situated.

The electrons within a SEM's beam hit the sample and interact with it to produce **secondary electrons, backscattered electrons** and x-rays, which are picked up by special detectors within the chamber. Secondary electrons are those ejected by the specimen itself due to interaction with the beam. Originating from the sample surface or a fraction of a millimetre beneath it, the number of secondary electrons reaching the detector affects the brightness of the signal: because the number of secondary electrons emitted is affected by the incidence of the beam to the sample surface, the image that results from the signal, once it is processed and sent to a display monitor, has a clear, three-dimensional appearance at magnifications of up to tens of thousands of times. Thus, very detailed morphological images of microscopic mineral crystals may be obtained. The operator may move the specimen around via controls situated at his or her monitor; thus several images of a sample may be obtained, or areas selected for analysis.

Backscattered electrons are high-energy electrons from the beam itself that have been reflected from the sample via interactions with the atoms making up its surface. Heavy elements are better at backscattering electrons, hence more electrons are collected, and in backscatter SEM mode, minerals made up of heavier elements will look much brighter than minerals made up of lighter elements; thus galena – lead sulphide – will appear bright white against quartz – silicon dioxide. Operators may switch between standard and backscatter modes in order to obtain as much information as possible (Fig. 5.7A, B).

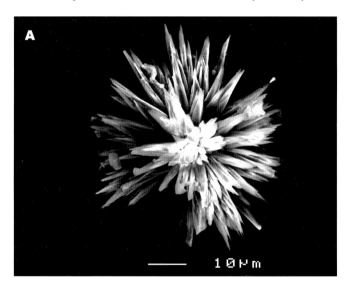

Figure 5.7 SEM images taken in (A) normal and (B) backscatter modes
(**A**) Electron microscopy makes it possible to examine very fine-grained mineral intergrowths. Here, a tiny (just 40 microns, or 0.04 mm in diameter) crystal cluster of an unknown scandium silicate mineral from North Wales (research in progress) dominates the field of view (**B**) Using an electron microscope in backscatter mode makes it possible to locate minerals or even areas within mineral grains containing heavier and lighter elements – the heavy elements stand out brightly. This backscatter image reveals complex chemical zonation (the stripy diagonal pattern) in a sectioned, 0.2 mm grain of manganocolumbite, a rare oxide of manganese, iron, niobium and tantalum, from Ireland.

X-rays produced when an electron beam hits a sample are useful because under such excitation, each element emits x-rays of a characteristic frequency: thus, the x-rays may be used for quantitative chemical analysis of mineral grains making up the surface of a polished section. There are two methods: wavelength-dispersive (WD) and energy-dispersive (ED) x-ray spectroscopy, which differ according to the way the x-rays are collected. Without going into too many technical details, EDS is quicker, but WDS is more precise. Collectively, this area of chemical investigation is referred to as electron probe micro-analysis (EPMA). Compositions of samples are determined by comparing the collected x-rays emitted by them against samples of precisely known composition known as standards.

5.5 Determining and describing parageneses

Parageneses of rocks or of mineral deposits are deduced by studying the relationships between the minerals that are present in order to determine the order of the events leading to their formation. In a very simple example, if concentric layers of minerals are found lining a fracture with a central, crystal-lined cavity, it will generally be the case that the oldest mineral in the sequence is that coating the fracture wall, with progressively younger minerals occurring towards the crystalline mineral lining the cavity, which will be the youngest. Often, however, repeated activation of a mineralized structure will result in earlier mineral assemblages being crosscut by veins of minerals belonging to later

assemblages. In some cases, repeated fracturing will be accompanied by brecciation, and clasts of older minerals will be cemented together by the newer minerals. Sometimes, one mineral will

be observed to have partially or totally replaced another, the chemical reaction front between the two marked by a series of concave bays in the mineral being replaced (Figs 5.8 and 5.9).

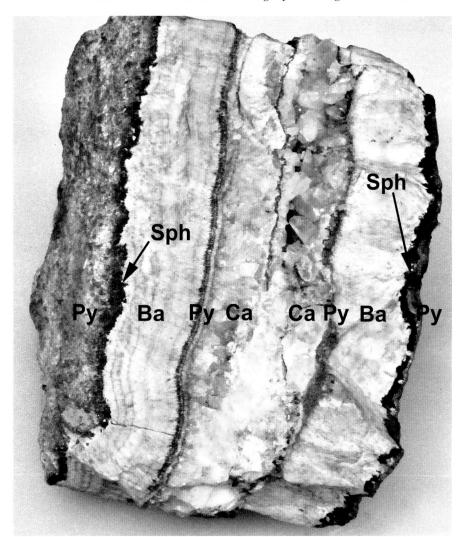

Figure 5.8 Paragenetic sequence

A simple paragenetic sequence in a thin (10 cm) vein. Minerals have clearly grown inwards from either wall: first to crystallize was pyrite (Py) forming slickencrysts along the walls. It was overgrown by sphalerite (black, Sph). Rhythmically banded barytocelestine (Ba) then started to precipitate until it had almost filled the vein; in the final, central cavity pyrite then overgrew the surface of the barite and the remainder of the cavity was part-filled with vuggy calcite

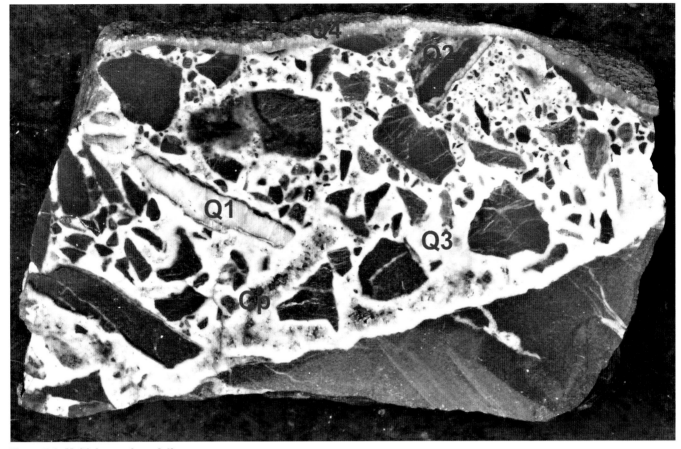

Figure 5.9 Multiphase rebrecciation

A more complex paragenesis is visible in this large sawn boulder from a lead-mine in Central Wales. Clasts, not only of the Ordovician mudstone wall-rock but also derived from the rebrecciation of earlier mineralization (Q1, late diagenetic milky quartz, and Q2, greyish quartz-cemented mudstone breccia), are cemented by a third generation of quartz (Q3) whose central vugs are lined with long prismatic crystals and patches of chalcopyrite (Cp). Finally, an open fracture has cut across the whole; grown upon this is yet another generation of quartz crystals (Q4). Therefore, four phases of mineralization, two involving brecciation, are represented in this single specimen.

However, things are sometimes more complicated, testing the researcher's arts of observation and deduction strenuously. If one mineral forms inclusions in another, logic ought to suggest that it formed either before or during the growth of the host crystal, and became enclosed. This is indeed mostly the case. A classic example is tetrahedrite $(Cu,Fe,Ag,Zn)_{12}Sb_4S_{13}$) inclusions occurring in galena (PbS), tetrahedrite is often highly argentiferous (silver-rich) and in many instances where lead-mining districts have a high silver content in their ores, it is found to be the microscopic culprit, invisible in hand specimens even under the hand-lens, but quite apparent once polished sections have been investigated. Other common inclusions found in galena are antimony- and arsenic-bearing sulphosalts and iron, nickel and cobalt sulphides and sulpharsenides/sulphantimonides (Fig. 5.10).

But there are instances where the apparently obvious is not always so. Inclusion-forming minerals may

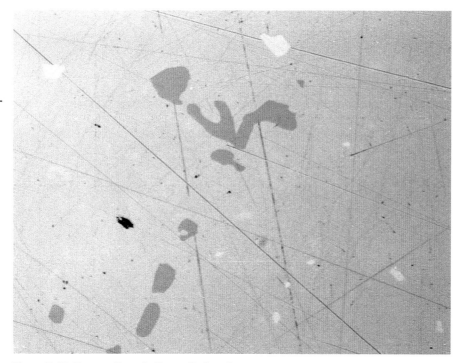

Figure 5.10 Bournonite and ullmannite in galena
A polished section of galena (field of view 0.6 mm) under high magnification revealing that it carries an assemblage of inclusion-forming minerals: anhedral bournonite (a lead-copper-antimony sulphosalt, dark grey) is accompanied by euhedral ullmannite (nickel sulphantimonide, white, and scratch-resistant due to its much greater hardness), showing that copper, nickel and antimony are also part of the overall mineralization.

such as **symplectites**, which look a bit like oil droplets in water, but are so fine that they can only be determined under very high magnification.

Then again, there are more complicated rocks and mineral deposits that have been recrystallized by metamorphism. Often, such changes are well documented and/or obvious at hand specimen or even outcrop scale. In areas with a history of deformation, for example, an important step is to establish whether structures such as intrusive dykes or quartz-sulphide lodes pre- or post-date the orogenic events. Examining outcrops will often yield the relevant information: the mineralogist will check for textural evidence such as folding or **boudinage** of a vein or dyke. Suspected deformation of a quartz vein can usually be confirmed by examining the quartz in thin section: quartz that has been affected by strain-related metamorphism typically recrystallizes into a granular mosaic, the grain-boundaries having a noticeably wavy (**sutured**) outline. In such an instance, where there exists clear evidence for quartz recrystallization, deducing the original paragenetic sequence of the associated minerals can be difficult because it is likely, unless evidence is found to show otherwise, that all of the minerals present have recrystallized, obliterating the original textures (Fig. 5.12A, B).

Some other common minerals are good indicators of deformation and recrystallization (Fig. 5.13A, B): pyrite, which is common in many rock-types, has a strong tendency to crystallize as euhedral cubes, and if its formation pre-dates deformation its crystals may exhibit diagnostic **pressure-fringes**. The latter are especially common in

develop through the process of exsolution. Exsolution occurs when a mineral, stable at one set of conditions, is subjected to a change – for example, cooling in an igneous rock, that takes it out of its stability field. It responds to such a change by unmixing into two or more minerals that are stable under the new conditions (Fig. 5.11A, B). Exsolution tends to occur in a relatively small and well-documented number of examples: a classic is the so-called perthite texture in feldspar, when the unmixing, during cooling, of a mixed sodic and potassic feldspar has

resulted in crystals of alkali feldspar, $KAlSi_3O_8$ containing parallel lamellae of albite, $NaAlSi_3O_8$. Likewise, in ore deposits with a high temperature of formation, chalcopyrite, $CuFeS_2$, often contains distinctive, star-shaped exsolution bodies of sphalerite, ZnS; their presence is quite diagnostic in terms of formational conditions. Some unmixing textures are very regular: for example, ilmenite may often be observed forming crystallographically-orientated networks of exsolution lamellae in magnetite. Others are irregular and can be very fine-grained,

A

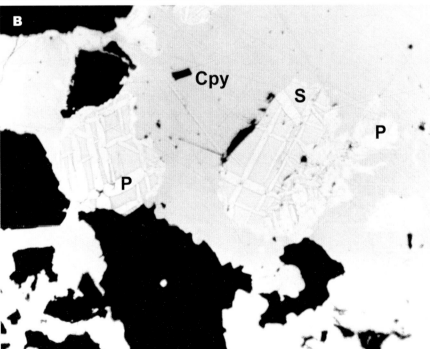

B

Cpy

S

P

P

Figure 5.11 Exsolution in polished section
In the Central Wales mining-district, the cobalt-nickel mineral siegenite is locally common. In this 10 × 6 cm hand-specimen, it forms pinkish-silvery metallic patches in quartz associated with chalcopyrite (**A**). It is not until the mineralization is examined in polished section (**B**) that it is realized that there are three minerals present: chalcopyrite ($CuFeS_2$) (Cpy), siegenite (($Ni,Co)_3S_4$) (S) and cobalt-rich pentlandite (($Co,Ni,Fe)_9S_8$) (P). The pentlandite forms a network of oriented lamellae in the siegenite and flamelike growths in the chalcopyrite, both textures strongly suggestive of exsolution. The individual siegenite crystals are up to 0.2 mm in diameter. This example again demonstrates the need to examine ore mineral assemblages in polished section in order to determine their precise composition and paragenesis.

slates: any pyrite crystals present act as relatively rigid bodies in a ductile matrix, during whose deformation quartz, chlorite and other minerals are dissolved from areas of high pressure, such as where the matrix is being compressed against the pyrite crystal, and are redeposited in the lower pressure zones on either side of the pyrite.

Another frequent feature of mineral assemblages that have recrystallized is the presence of porphyroblasts. A porphyroblast is a large, often euhedral, crystal of any mineral that has grown within the finer-grained groundmass of a mineral assemblage: commonly found in metamorphic rocks, they also occur in recrystallized ore-deposits. In both cases the process involves the diffusion of various ions to preferential crystallization sites, where, as they grow, they may enclose grains of

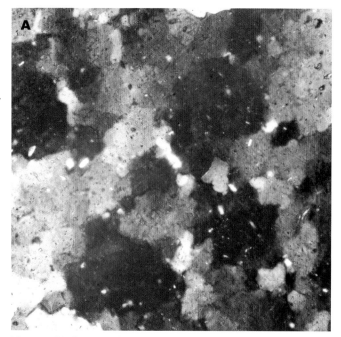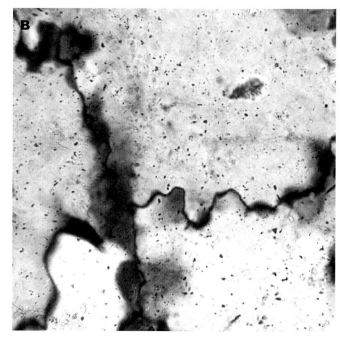

Figure 5.12 Quartz deformation

Two examples of vein-quartz from North Wales that has recrystallized during regional, compressive deformation, both viewed in thin section under crossed polars. The first image (**A**) shows a granular texture with grains not going into and out of extinction clearly, but instead having a variably shadowy appearance. The second image (**B**), zoomed in (field of view 0.18 mm) shows the highly sutured appearance of grain–grain boundaries: these are both diagnostic features of quartz that has recrystallized under strain.

other minerals in their growth positions, and thereby preserve existing fabrics. As such, they can be useful aids in determining a rock's history of deformation and metamorphism.

When examining rock or ore samples, the distinction also needs to be made between primary and secondary parageneses. All rocks and ores, when exposed to the elements, weather to varying degrees, during which process elements of the primary paragenesis are removed and new secondary minerals deposited. Generally, the products of weathering, such as crusts of loose, rusty iron oxides, are obviously thus formed; however, in some cases care is required to distinguish

primary and secondary minerals. The copper iron sulphide bornite (Cu_5FeS_4) is an example of a mineral that can form in both primary and secondary parageneses, and in other similar cases experience is one's best ally in discerning the actual relationship.

Once a paragenetic sequence for a rock or a mineral deposit has been discerned, it is expressed both as a written petrological description, where the textural relationships of the various minerals are described, and as a **paragenetic diagram**. The latter allows the reader to appreciate the paragenetic sequence at a glance. Suppose, for example, that you have a mineral deposit in which argentiferous galena

is found to contain tetrahedrite inclusions, is overgrown by quartz, and then both argentiferous galena and quartz are fractured and veined by sphalerite containing inclusions of pyrite. The paragenetic diagram expressing the sequence would look something like this:

```
galena        ---------------
tetrahedrite    ---
quartz               ----------------

xxxxxxxxx FRACTURING xxxxxxxxxxxxxx

pyrite                          --
sphalerite                      -----------
```

It is generally the case that, once the paragenetic sequence of an individual

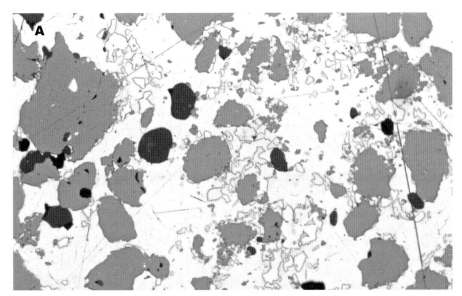

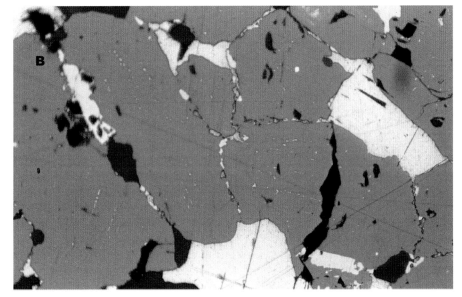

Figure 5.13 Recrystallized sulphides
Two examples of deformation textures in polished sections of massive intergrown sulphides from a known compressively deformed vein in North Wales (it exhibits strong boudinage at its outcrop). Sphalerite (dark grey) has recrystallized and reveals a rather granular texture, grain margins being picked out in the second image (**B**) by concentrations of chalcopyrite (yellow). In the first image (**A**) there is also much pyrrhotite (pinkish-white with relief) that appears to have responded to the deformation by shattering. The pale, greyish-white matrix mineral is in both cases galena, which in hand-specimen reveals textures consistent with deformation by flow, a common observation with this soft (2.5 on Mohs scale), easily cleaved mineral. Establishing the original paragenetic sequence of this material would involve a great deal of painstaking work. Fields of view are 2.5 and 1.25 mm respectively.

5.6 Further studies

When, and only when, the detailed paragenetic sequence and structure of a mineral deposit has been established, more specialized studies may be undertaken. It is beyond the scope of this book to go into them in detail, but they include:

◆ **Fluid-inclusion studies** – examination of drops of the original hydrothermal fluid that may be present and observable in translucent minerals such as quartz and that can yield up data that allow estimates of physical conditions at the time of mineralization (temperature and pressure) and give indications of the chemistry and thereby the type of hydrothermal fluids that deposited the mineralization;

◆ **Isotopic studies** – certain minerals contain stable and non-stable that

rock unit or mineral deposit has been established, it will hold good in any sample taken from there, unless errors have been made in the determination of events. Errors are most likely to occur because of limited sampling missing some aspect or other of the minerals and their relationships, and the more complex a mineral deposit is, the more care has to be taken during the sampling procedure.

permit their approximate age to be determined (Pb-Pb in galena; U-Pb in zircon; K-Ar and Ar-Ar in mica; Rb-Sr in feldspars and other silicates; Sm-Nd in silicates, to name a few commonly-used examples). Isotopic dating can be a reliable method, but caution is required at the same time: for example, in K-Ar dating, argon-loss can be a problem if the mineral deposit has been disturbed post-formation by deformation, allowing argon to escape from the system.

◆ Isotopic studies that give information with respect to the source of the mineralizing fluids: a well-known technique involves looking at sulphur isotope ratios in sulphides, which can indicate whether the fluid had a magmatic origin or not.

◆ Minerals that act as **geothermometers** or **geobarometers** – certain minerals occurring together,

or one mineral that has exsolved from another can indicate whether a mineral assemblage has formed at high or low temperatures or pressures.

◆ **Trace-element geochemistry** – levels of certain rare elements in common minerals may give information as to the origin and/or affinities of a rock unit or mineral deposit.

All of these techniques have one thing in common, though: the results *cannot* be properly interpreted unless one is certain of the paragenetic sequence and structure of a rock unit or mineral deposit. To give a made-up but entirely plausible example: an orefield has two generations of mineralization consisting of high-temperature gold-bearing (auriferous) veins and low-temperature non-auriferous veins. The auriferous veins (stage A) are of Silurian age and the non-auriferous veins (stage B) are of Carboniferous age, and

they cut and brecciate the auriferous veins. Both sets of veins contain much galena, sphalerite and chalcopyrite and because both sets were encountered during mining, mine-tips contain lumps of sulphides from them both. It is clear that without first determining the paragenetic sequence, a single, randomly-grabbed chunk of galena from a mine-tip could give a Pb-Pb isotopic date that says that the mineralization is, depending on pure luck, of either Silurian or Carboniferous age. However, by undertaking paragenetic studies and determining that there are two stages of mineralization, Stage A and Stage B can then be sampled in an informed manner and the data they yield will give the researcher a far better chance of drawing robust conclusions as to the true sequence and nature of mineralizing events.

Introduction

This chapter covers what is generally known as the **Minerals Industry**. In industrial parlance, the term 'minerals' refers to any solids, be it individual minerals (such as galena, mined as the major source of lead) or intergrown minerals (such as granite, quarried for a range of purposes from ornamental stone to road-aggregate) that can be extracted from the ground economically. In other words, it refers to things that may be mined or quarried at profit. The industry is diverse – a reflection of the diversity of minerals that are present in the Earth's crust. Mineral extraction is undertaken for four key purposes. Firstly there are the **ores**: minerals from which metals and other useful elements are obtained. Secondly, there are the **industrial minerals**, which, owing to some important physical or chemical property, are useful in a variety of industrial processes. Thirdly, there are the **chemical feedstocks**, used for the manufacture of chemicals such as mineral fertilizers for agriculture; and fourthly there are the minerals that are obtained purely for their aesthetic value.

6.1 Minerals extracted from ore deposits

In **economic geology** terms, an ore-deposit is one which is economically viable to work. Potential ore-deposits are assessed in terms of two key parameters plus a number of peripheral ones.

Once a deposit has been discovered, it is systematically explored in order to determine (a) its **grade** and (b) its **tonnage**. The process, usually involving the drilling, recovery and analysis of regularly-spaced cores, and in some cases the driving of exploratory shafts and tunnels, is basically designed to give the best estimate of how many tonnes of mineralized rock there are, and what overall percentage of metals is present.

One of the peripheral factors kicks straight in at this point, and that is the current and forecast price(s) of the metals on the global market. Low-grade ore-deposits are especially vulnerable to volatile prices, but if prices are high and stable they may be viable. In periods of price volatility, some parts of an existing mine may be mothballed because the percentage of metals in the rock fall below the break-even point (or **cut-off grade**), but mining in such areas may resume once prices have risen once more. At some points in the past, combinations of events have depressed prices so badly that entire **orefields** have closed down. This happened in the United Kingdom after the First World War; in several orefields there was a thriving zinc-mining industry, but two things happened: firstly the release of Government stocks of zinc and secondly the advent of the ore-processing technology known as **froth flotation**. The latter allowed huge waste-heaps rich in previously unextractable zinc

minerals at overseas localities such as Broken Hill in Australia to be efficiently processed. The market became flooded with zinc, the price went through the floor and the UK mines largely became uneconomical and closed.

Other peripheral factors include accessibility: the cost of opening up a mine increases with its remoteness from existing infrastructure. Political stability is a major factor in some parts of the world. Environmental legislation varies from country to country: although in an ideal world any mining company would seek to have a very good environmental track record, this is sadly not always the case. Climate may affect the viability of operations, especially at high altitudes or latitudes where seasonal shutdowns may occur.

These factors all apply to both industrial minerals and chemical feedstocks. In addition, the techniques of looking for economic deposits of all three categories are broadly similar, so it is appropriate to include a quick primer about mineral exploration at this point.

A company (or government agency) involved with the minerals sector will first make a **desktop study** of a region in terms of the available geological literature and using **remote-sensing** techniques, the key question being, 'does the geology of this area suggest that mineral deposits may be present?'. If the answer is yes, then the next phase begins: the ground work. Firstly, access needs to be arranged. This can be a time-consuming process. It is easier where everything is state-controlled and more complicated where ownership is divided. In the UK, for example, as a consequence of history, anyone wanting to look for gold needs first to take out a Prospecting Licence from the Crown Estates, who own precious metals over pretty much the whole country. Then it is necessary to identify land surface ownership and make approaches to the surface owners for permission to prospect, which may or may not be granted. Often a legal agreement will have to be drawn up and the landowner compensated financially – and that is before anything has been found!

Once access is arranged the work can begin. One of the quickest ways to evaluate a large region, maybe hundreds of square kilometres in extent, is to do a **drainage survey** or, if a government agency has already done one, to buy the data. Drainage surveys involve sampling stream sediments. Since these are – with the exception of areas that were glaciated during the Quaternary – derived from the local rocks, they should entirely represent the geology of the immediate area, including any mineralization. In glaciated terrain, the situation can be more complicated due to the possibility of minerals having been introduced by ice-sheets from far away, but the problem can be overcome by careful observation and interpretation of data, including the mapping of any glacial drift and noting the abundance of exotic clasts.

Drainage surveys vary in their thoroughness from the quick reconnaissance, one sample per tributary approach to more detailed programmes

in which an entire catchment is sampled in a regularly-spaced manner with the distance between sampling-points being perhaps 100 m or less. The former approach would be taken if examining a large area not known for mineralization on a moderate budget; in areas where mineralization is already known to occur, the latter approach, though more expensive, may be justified. Either way, two samples are typically collected from each locality: a **fine-fraction stream sediment** sample and a **heavy panned concentrate**. Sampling sites are chosen carefully in order to avoid any obvious signs of manmade contamination such as roads, buildings or rubbish dumped nearby.

A typical sampling procedure at a chosen site might be as follows: two pans and two sieves are firstly rinsed downstream from the site to avoid sample cross-contamination. Next, one pan is placed on a flat spot on the riverbank. A fine (150 micron mesh is commonly used) sieve is placed on top of that pan, and on top of it in turn is placed a coarser (2 mm mesh) sieve. The sampler then digs a hole into the river-bed, placing the gravels, silts and so on into the top sieve. Sediment that will pass through that sieve is washed through, the remaining oversized fraction being discarded after noting the types of clasts present. When enough material has collected in the lower sieve, rubbing by hand forces the fine fraction through into the pan beneath. Once there is enough fine material for analysis (about 100 grams) the pan is left to settle, the water is carefully drained off and the sample is bagged, sealed and labelled. The remaining <2 mm material can be panned in the

second pan whilst the fine sample is settling, to provide a similar amount of heavy concentrate. The concentrate is examined for any potentially interesting minerals (such as grains of gold), bagged, sealed and labelled. Any relevant observations are noted, and the grid reference and GPS position of the site are logged.

Once the samples have been collected, they are sent off for preparation and chemical analysis for a wide range of elements. These may not only include the target metals such as gold, but also so-called indicator or **'pathfinder' elements**: many gold deposits also contain a lot of arsenic and/or bismuth minerals, so that high levels of either of these elements in the stream sediments of part of a Licence area would make it worthy of more detailed examination.

Once the results are back from the lab and plotted on maps, an overview of the geochemistry of the area is possible, and areas with anomalous (higher than background) levels of target and pathfinder elements can be chosen for more detailed work. Such areas are then walked over carefully, the geologist looking out for any mineralized outcrops or loose blocks that allow closer examination. Some areas that feature no, or very poor, outcrop, are taken straight to the next stage – a grid is surveyed out and marked, and regularly-spaced samples of soil and/or deep overburden are taken, perhaps every 10 or 20 metres along each gridline. Again these are bagged, sealed and labelled and sent off for analysis. Based on the results the area is then either written-off or taken to the next stage, which involves seeking to locate and directly sample the mineralization and any buried mineralized structures. If the overburden is not too thick, strong geochemical anomalies may be investigated by mechanized trenching; any mineralization located can then be sampled and analysed, both in terms of its geochemistry and its mineralogy. In areas of deep overburden, geophysical techniques are used, working along the same grid. Geophysics can detect structures such as faults and buried concentrations of conductive or magnetic minerals.

If the geochemical and geophysical results confirm the interest in an area, the next stage is to investigate the three-dimensional properties of the mineralization by **diamond drilling**, by far the most expensive part of the operation so far, so that it needs to be strongly justified. Often, the first stage involves a few exploratory holes targeting specific geochemical and/or geophysical anomalies, drilled to test what is present and to confirm in broad terms that it continues with depth. The procedure is the same in all cases: the core is recovered and laid out in wooden core-boxes with depths annotated and sections with poor recovery noted. It is then sawn in half lengthwise, and its entire length is logged by the geologist. Sections of interest are sent for analysis (retaining the other sawn half for reference). If, and only if, the analytical results are satisfactory, a regular series of boreholes is then done. Once these are complete and the data are in, the exploration team can model the deposit using software designed for the purpose and arrive at a good idea of how many tonnes they have, what minerals are present and at what grade. The decision as to whether to go ahead and plan mining operations can then be made (see Fig. 6.1).

Figure 6.1 Diamond drilling
Diamond drilling at a gold prospect in Scotland: emptying the core-barrel into a core-box, which allows the drill core to be transported, stored and examined without its running-order down the whole getting mixed up.

Mineralogical studies become increasingly important with every stage as outlined above, and especially when considering whether or not to mine. Using gold as an example again, the nature of its occurrence makes a real difference to the planning process. Gold deposits vary a great deal. In some, the gold is coarse, meaning it is easily visible to the naked eye and its extraction may be done by simply crushing the ore to a fine powder and using water to wash away the lighter waste materials. If such '**free milling gold**' occurs with very little sulphide in a quartz matrix, there will be fewer problems with respect to the waste - the **tailings**. However, if the gold occurs in solid pyrite, then the latter can lead to acid mine drainage issues as it oxidizes. If the iron arsenic sulphide, arsenopyrite (FeAsS), is present (which it often is) then there is another toxin to worry about. In other deposits, the gold is so fine-grained that it can only be seen with a petrographical microscope, even though the geochemical grade is impressive. That will mean that chemical extraction (such as cyanidation) may be required, changing the cost of the whole operation. The mineralogist therefore has to study the deposit, unravel its petrology and paragenetic sequence and identify minerals, their associations and grain-sizes. In doing so, critical information is unearthed that will largely dictate the **planning stage** (Fig. 6.2).

Planning is a complex process that will vary from area to area depending on the regulations that are in place; these may concern types of permitted development and environmental impact of an operation, for example. At this stage the mineralogists and geologists hand over the proverbial baton to solicitors and planning advisers for what can be a time-consuming and expensive process. There may be strongly-held objections to face; it is a fact of life that whilst people freely make use of all sorts of things made from extracted minerals in their everyday lives, they tend to prefer them to be sourced elsewhere, where someone else has to look at the operation from their windows.

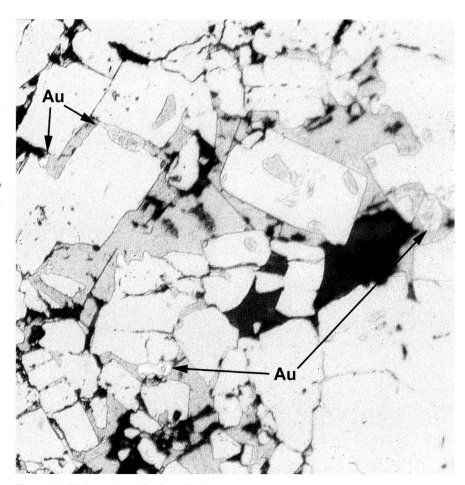

Figure 6.2 High-grade gold mineralisation
A photomicrograph of a polished section of gold-bearing sulphides from a prospect in Scotland. The gold (yellow) occurs, together with galena (grey), overgrowing earlier pyrite (yellowish-white). The biggest grain of gold in this field of view is just 40 microns across. Other grains are still smaller. In mine-planning, consideration of gold recovery techniques would need to take such factors into account; very fine gold can be difficult to recover efficiently using gravity alone.

If planning permission is obtained and the mine is finally up and running, an ore-deposit tends to be worked in a series of blocks of known grade and tonnage. This helps mine planning and furthermore can be advantageous if commodity prices swing about. Areas that are worked-out are often backfilled with waste rock (known as 'deads'), which solves two problems: it helps with ground stability and it is somewhere to dump material that would otherwise need to be tipped at surface. The ore is then brought to a mill, situated either at surface or within the mine, where it is crushed and the materials of value extracted using a variety of technologies. It is then generally sent off-site, as an ore concentrate, to be smelted. Concentrates may consist of a single ore mineral or a mixture, depending on the mineralogy of the deposit, and different techniques may be required in order to extract the metals they contain. Again, such factors are of considerable importance in determining the economics of an operation, and it is the mineralogist who will identify them on an ongoing basis.

Most metals occur in a range of minerals, but generally only a few are common enough to be of economic interest. For example, most lead is obtained from galena; in the case of zinc, sphalerite (ZnS) is the key source. Copper, though, may be obtained from minerals like chalcopyrite ($CuFeS_2$), bornite (Cu_5FeS_4), chalcocite (Cu_2S) and more complex minerals like tetrahedrite $(Cu,Fe,Ag,Zn)_{12}Sb_4S_{13}$. Tetrahedrite is a widespread mineral that exhibits a rather variable composition from deposit to deposit; at some localities it is very rich in silver, for example, while other tetrahedrites can be silver-poor. Silver is also obtained from a wide range of other complex minerals and is worked in numerous places as the native metal, as is gold. Gold also occurs, in some cases to an important extent, in compounds with bismuth, antimony and tellurium; some rich gold deposits consist almost entirely of gold-bearing tellurides.

Some rarer metals, although they form minerals, tend to be produced commercially as by-products from other ores. Cadmium is a good example: its sulphide, greenockite (CdS), is rare but the metal readily substitutes into the lattice structure of sphalerite so that most world production is recovered as a by-product of zinc mining and smelting. Although silver is mined in its own right, it may likewise be present in galena, and it has long been a by-product of lead mining in many parts of the world. Indeed, mineral deposits producing more than one metal are the rule as opposed to the exception. Another example is tin and tungsten: the respective ores, cassiterite (SnO_2) and wolframite $(Fe,Mn)WO_4$ often occur together, especially in granite-hosted vein deposits. Sphalerite and galena are very commonly found together, with mines producing zinc, lead and by-product silver. Often in association with galena and sphalerite are baryte ($BaSO_4$) and fluorite (CaF_2), which are industrial minerals, bringing us to the next section.

6.2 Industrial minerals and chemical feedstocks

Many minerals occurring in nature are not ores as such, but because of their chemistry and/or physical properties, are of key importance in our modern industrialized society. An exhaustive consideration of all the industrial minerals and chemical feedstocks would require a whole book dedicated solely to that subject, so instead we will take a look at some common examples and why they are important.

Baryte, barium sulphate ($BaSO_4$), is a pale-coloured mineral that crystallizes in the orthorhombic system, often forming aggregates of platy crystals with a vitreous to pearly lustre. It is soft, with a hardness of 3–3.5 on the Mohs Scale, but two things make it stand out: its chemical inertness and its high density – $4.5g/cm^3$, or almost twice that of quartz.

Baryte has a number of industrial uses, such as fillers in paint and plastics, in radiation-shielding cement and in ceramics. However, almost 80% of worldwide production is for use as **drilling-mud**. As a consequence of its high density, baryte is an excellent weighting agent for drilling fluids in oil and gas exploration, when holes are often drilled into high-pressure environments where blow-outs are a risk. Baryte used for drilling such wells can be black, brown, pink or grey – a property that varies from deposit to deposit; unlike its role in paints and plastics, the colour is unimportant. For drilling-fluids, the baryte is finely ground down to a few tens of microns; the mineral must be clean so that the required density value is maintained and it is free from contaminants such as quartz, whose hardness may otherwise cause unwanted abrasion to the expensive bearings of the drill-bit. It is a classic example of an industrial mineral whose usefulness is a direct result of its physical and chemical properties.

Baryte is often obtained as a by-product from lead-zinc mines, but it also occurs in deposits where it is the dominant mineral species. The same observation applies to that other common industrial mineral, fluorite (CaF_2). The name 'fluorite' is derived from the Latin, 'fluere', meaning 'to flow'. In this case, the important property is not the density, but instead that when mixed in with the charge that goes into a smelter (ore and fuel) it lowers the melting point of the waste slag, making it runnier and therefore easier to draw off: it is, in other words, a **smelting flux**.

The material used as flux is termed 'metallurgical grade fluorite', which is the lowest commercial grade at 60–85% CaF$_2$. The next grade up, 'ceramic grade fluorite', at 85–95% CaF$_2$, is used in the manufacture of opalescent glass, enamels and cooking utensils. The highest bulk commercial grade is referred to as 'acid grade fluorite', at 97% or more CaF$_2$. This is a chemical feedstock, by which it is meant that it is used to produce another chemical, hydrogen fluoride, an important catalyst used widely within the petrochemical industry and the starting point for the manufacture of hydrofluoric acid.

The hardness or softness of some minerals makes them of major importance in industrial terms. Take diamond, for example. For every gem-quality diamond that is mined, there are many more of poor quality; in fact some diamond deposits consist almost entirely of non-gem quality stones. However, because of its exceptional hardness (10 on the Mohs Scale), diamond deposits that are devoid of gemstones can still be worth working because that great hardness makes it useful as a strong abrasive. A steel circular saw-blade, with its rim consisting of diamond dust embedded in a tough solder alloy, will slice through a block of quartz in no time. An even more important use is in drill-bit manufacturing, hence the term 'diamond drilling', which one often encounters in mineral exploration reports – it does not mean that a company is drilling in search of diamonds but that it is using state-of-the-art drilling equipment in its evaluation of an underground mineral deposit it has discovered.

Softness, one might think, is a less useful property, but there are several soft minerals that are extremely useful. Graphite, powdered and mixed with varying proportions of clay, provides pencil-leads of varying degrees of hardness or softness. The softness of the molybdenum sulphide, molybdenite (MoS$_2$), in combination with its perfect basal cleavage, makes it of great use as a lubricant. This is entirely down to the mineral's chemical structure, consisting of lamellar plates in which each molybdenum atom is sandwiched between two sulphur atoms. As the manufacturer of the lubricant Molyslip explains:

The sulphur atoms are attracted to metal and therefore become plated or bonded on to each of the adjacent bearing surfaces. In between these two platings further layers of molecules form. The sulphur-to-metal bonding is very strong, but the sulphur-to-sulphur bonding between adjacent molecules is very weak. Thus, there are two bearing surfaces, each protectively plated by a layer of molybdenum disulphide molecules with sliding or lubricating layers of molecules in between. In this way direct contact of metal-to-metal surfaces is prevented, friction is considerably reduced, with the consequent elimination of local heating, wear is inhibited and protection achieved even under extreme conditions of pressure and temperature.

The molecular thickness of molybdenum disulphide is such that there are approximately 40,000 lubricating or cleavage planes in an MoS2 film one thousandth of an inch thick! The molybdenum disulphide plating is, in effect, a separating layer of immense strength, greater than the yield stress of most metals...and in addition it possesses the low coefficient of friction of .03 to .06 which gives more efficient lubrication combined with this greater protection.

Tolerance to high-temperature environments is another useful property shared by a number of minerals, which are referred to as being **refractory**. These substances can withstand temperatures well in excess of 500 °C without softening or otherwise deforming, so they have many industrial applications, from large furnaces and kilns down to laboratory crucibles. Many of the naturally-occurring aluminosilicate minerals have refractory properties, as do the oxides of calcium, magnesium and aluminium. One of the most refractory minerals is zircon (ZrSiO$_4$), the melting point of which is about 2200 °C, which means it is so indestructible that it can even survive partial melting of its host rock deep in the Earth's crust, so that a granite that is, say, 500 million years old may contain 'inherited' zircons that are twice that age. It is easy to see the potential uses of such durable minerals when one thinks of some of the extremes into which human endeavour has led: the tiles on the underside of the space-shuttle are just one obvious example of something that would be impossible without refractory minerals and the more specialized substances manufactured from them.

Some minerals that form large crystals of high clarity have their uses in optical equipment. Here we return briefly to fluorite, because it is one of the most important examples of this kind: it has a very low refractive index and **dispersion**. The latter, when high, causes the separation of white light into components of different wavelengths – a good example being a rainbow, where the result is a spatially arranged colour spectrum of visible light. In optics, dispersion is bad news, so a mineral that has low dispersion is potentially useful. At one time, lenses were made from so-called optical grade fluorite; today, fluorite lenses are manufactured from high-purity fluorite that has been melted and combined with other materials. Cameras, microscopes and telescopes all benefit from the properties of this mineral.

Some industrial minerals are mined in vast quantities as bulk chemical feedstocks with everyday uses. A good example is potash – a collective term for water-soluble potassium salts such as sylvite (KCl), which is an important component of some evaporite deposits. Other compounds occurring in evaporite deposits, such as halite (NaCl), are important both as feedstocks and as industrial minerals (road-grit is but one use, another being, when refined, table-salt). Gypsum is common in many evaporite sequences and is the raw material for the manufacture of plaster-of-paris. The process involves heating the mineral to about 150 °C, thereby driving off 75% of its water of crystallization:

$$4CaSO_4 \cdot 4H_2O = 4CaSO_4 \cdot H_2O + 3H_2O$$

This reaction is reversible: when the dry plaster powder is mixed with water, it recombines, forming gypsum once again, which starts to harden within a few minutes, something to which any plasterer who has left a dirty trowel uncleaned during their lunch-break will testify. In the case of another common evaporite mineral, celestine ($SrSO_4$), the main use is as

a chemical feedstock: celestine is the main source of strontium, which among other things, produces the red colour often seen in fireworks when its nitrate is included in the 'recipe'.

Calcium phosphate minerals of the apatite group ($Ca_5(PO_4)_3(F,Cl,OH,CO_3)$) can occur – usually in a cryptocrystalline form – in bulk in some sedimentary rocks: known as **phosphorites**, which are mined and processed in huge quantities to make inorganic fertilizers. And of course there is calcium carbonate – calcite, the bulk feedstock for cement manufacture. Modern cement-making is a high-tech business because of the demand for cements that can be used in exacting environments, but the original method was much simpler:

$$CaCO3 = CaO + CO2$$

All that was necessary for that reaction to take place was to quarry some good limestone (calcium carbonate), crush and powder it, heat it in a kiln to between 800 and 1000 °C for a day or so, and you would be left with quicklime (calcium oxide). That would then be transported to where it was needed and mixed with water to form, in a strongly exothermic reaction, slaked lime, $Ca(OH)_2$. The important bit then happens: once slaked lime dries out, it begins to react with atmospheric carbon dioxide:

$$Ca(OH)_2 + CO_2 = CaCO_3 + H_2O$$

This is how traditional lime mortar was made and how it 'went off', hardening over a matter of a few days.

The colours and other physical properties of various minerals also make them useful in ceramics, as glazes, which are used for a number of reasons: importantly they seal off the porous surface of pottery, making a

surface safe to eat or drink from, and this surface may, according to the ingredients used, have variance in colour, lustre and texture for ornamental purposes. Many glazes are based on silica – quartz – but they must include a flux, which enables the glaze mixture to melt at a certain temperature, a stabilizer that stops all the glaze simply running off when it melts, and small amounts of additives to change the opacity or give various colours. A wide variety of minerals are used for these purposes.

As well as silica, the feldspars are often used as a glaze-base: fluxes include dolomite ($CaMg(CO_3)_2$) magnesite ($MgCO_3$) and witherite ($BaCO_3$). Kaolinite ($Al_2Si_2O_5(OH)_4$) is commonly used as a stabilizer. Recipes for glazes are determined by the intended appearance: for example, the ratio of silica to alumina in a glaze determines the lustre of the finished piece, so that a glaze with a silica–alumina ratio of 9:1 will result in a shiny surface, whereas ratios of 4:1 and less would result in a matte finish. Likewise, the amount of flux is critical: too little and not all of the silica will melt, leaving a matte surface. Colour is typically added by utilizing small amounts of various transition-metal (or sometimes rare-earth) oxides.

6.3 Ornamental/aesthetic uses

The mineral specimen market has already been covered in chapter 4, but that deals entirely with unadulterated natural crystals. There also exists a major global market in minerals that, through careful preparation, have lost their original physical form but as a consequence are particularly showy in appearance: in other words, the **precious** and **semi-precious** gemstones.

The classification of gemstones goes back to the time of the ancient Greeks; it is therefore more of a tradition than an empirically-based division. However, the gemstones that are widely classified as precious have a number of things in common: great hardness, high lustre and great rarity being at the forefront, diamond being a classic example. In addition, ruby and sapphire, which are red and blue variants of corundum (Al_2O_3) and emerald, the transparent, deep green variety of beryl ($Be_3Al_2(SiO_3)_6$), have been traditionally regarded as precious gemstones, whereas all other gem-quality minerals have been bunched together as semi-precious. However, this classification is not necessarily reflected in prices, and it is fair to say that the precious/semi-precious division is rather misleading; some minerals that are (a) rare and (b) even rarer in transparent and flawless crystals are, when faceted, of a similar price-range to the precious stones. A recent browse of a gem dealer's website demonstrated this point: a good Colombian emerald of 0.88 carats (1 carat = 200 mg or 0.2 g) was offered at $850, whilst a beautiful deep green tsavorite (a rare form of garnet) of 0.89 carats was $650.

The optical properties of minerals are what determine, when transparent and flawless, their appearance when **faceted** and polished. These properties include refractive index, lustre, dispersion and pleochroism. Cleavage can be important, as cleavage planes may show up as internal **flaws**: inclusions may be regarded as either good or bad. Large fluid inclusions may add interest to a stone, and of course polished rutilated quartz – clear rock crystal full of lustrous, golden needle-like crystals of rutile (TiO_2) – is a gemstone in its own right. But in general, the more flawless and inclusion-free a stone, the greater its value. Diamonds are graded by what are known as the '4 C's': colour, clarity, carat weight and cut. The latter is especially important with colourless diamonds because the main attraction (to some people, anyway) is that they sparkle so brilliantly. In comparison, rough diamonds straight from the mine, even when well crystallized, have a relatively dull appearance. The aim, therefore, of cutting is to produce a faceted jewel in which the angles between the facets optimize the lustre of the stone – which is determined by the optical dispersion of white light. At the same time, the number and total surface area of facets would determine the weight of the product – and the amount of wastage involved in the cutting process (Fig. 6.3).

Cutting of diamonds – or any other gemstones – is a painstaking and delicate process. Specialized tools are required and above all, experience: although diamond is extremely hard it is also a brittle mineral with a perfect cleavage, so there is a lot that can potentially go wrong. An experienced

Figure 6.3 Gemstones
A selection of cut precious and semi-precious gemstones, illustrating just a small percentage of the diversity of types of cut that are used.

cutter will take into consideration the final intention for the finished stone (and there exists a wide range of 'cuts' or faceting-patterns to choose from), and then plan the cutting meticulously. Given the high value of exceptional diamonds, this is not a job to rush at: it may take weeks or even months of careful examination of a high-end rough stone before committing to the actual physical work. The assessment of such a stone would involve an investigation into its crystallography – many rough diamonds are not perfect crystals, so that study is needed in order to work out where its cleavage planes lie. Flaws need to be detected, likewise inclusions of other minerals.

The initial steps in the cutting process involve trimming the stone down to that which will be faceted and polished. This is a risky process; originally done by a heart-stopping hammer-blow to a pointed metal tool, today diamond saws and lasers are also used. Once that stage is over with, the slow part of the process commences – the forming of each polished facet on the diamond, which involves setting up the equipment to work at precise angles as spinning, industrial-grade diamond-coated plates grind each facet down to its specified depth. Polishing these ground facets finishes the job off; at this point, the stones will be inspected for any flaws that were missed earlier in the process and graded accordingly. Flaws may not be too big an issue if the stone is to be set in metal: it may be arranged so that the metal conceals the flaw.

In the case of coloured gemstones – including diamonds – it is the rarity of the colour (for example, blue diamonds are regarded as especially prized) and

its intensity, notwithstanding other common properties such as lack of flaws or inclusions that are detrimental to appearance. The natural colour is primarily down to the chemical composition of the gemstone: that determines what part of the daylight spectrum of visible light is either absorbed or reflected by the stone. Emerald, for example, absorbs red and blue light whilst reflecting green – hence its green appearance. In its pure form, beryl is colourless; however, a small amount of impurity in the form of chromium gives the emerald-green colour. Beryl can carry a range of impurities – the presence of manganese gives it a pink colour (variety morganite) whilst ferrous iron gives it the beautiful blue-green that is displayed by the variety aquamarine.

As well as colour, some stones are valued because of the optical effects they display, a well-known example being star sapphires, which when cut and polished display, beneath an overhead light-source, a pale, starlike structure set against the blue background, generally six-rayed but sometimes twelve-rayed. The effect is due to inclusions of other minerals: in a well-developed crystal, it is not uncommon to find inclusion-forming minerals present as symmetrically-orientated intergrowths. In the case of star sapphires, it is needles of rutile that do the trick, sometimes accompanied by tiny platelets of haematite. Such stones are cut '**en cabochon**': that is, instead of faceting, they have a selected area worked into a smooth, polished dome. Selection is important because of the orientation of the inclusions relative to the sapphire's crystal axes. If the cabochon face is parallel to

the crystal's c-axis then the star should show well at the top of the dome. Get that orientation off by a few degrees and the star will appear off-centre.

Other stones cut into cabochon domes include opal and tiger's-eye, both of importance primarily for their plays of light and colour, rather than the colour and clarity by which the commonly-faceted stones are valued. However, in the gemstone world, there are two other factors to be aware of: these are that naturally-occurring stones may be treated to 'improve' their colour, and that some varieties may be created artificially. Here's an example: citrine – clear yellow crystalline quartz – is actually a rather uncommon variety of the mineral in nature; the purple variety amethyst is much more widespread. But if crystals of amethyst are strongly heated, altering the chemical state of the transition metal impurities that give them their purple colour, they turn yellow and stay that way upon cooling. The 'citrine' that is commonly found next to the dyed blue slices of agate in gift-shops is almost always heat-treated low-grade amethyst. In a similar way, clear rock-crystal, if irradiated, converts into smoky quartz.

It is often said that if something walks like a duck and quacks, then it is probably a duck. Not necessarily so in gemmology: step forward **cubic zirconia** (CZ). This is synthesized, crystalline zirconium dioxide (ZrO_2), a substance that occurs but rarely in nature as the mineral baddeleyite, discovered in the late nineteenth century. Synthesis of fine-grained zirconium dioxide, useful for its highly refractory properties, was widespread by the mid–twentieth century, but it was only in the 1970s that the production of large crystals was accomplished by scientists in the former Soviet Union; once that point was reached, production for the gemstone market began in earnest.

Cubic zirconia is hard, optically flawless and usually colourless, with an adamantine lustre, although additives will produce versions in various colours, just as natural impurities give the various coloured varieties of beryl. It has a high refractive index, and dispersion is greater than that of diamond. In other words it is a great diamond-simulant. However, there are ways of distinguishing the two. CZ is softer (8.5 on the Mohs Scale) and it is a lot denser at 5.68 g/cm^3 as opposed to 3.53 g/cm^3 for diamond – a one carat faceted CZ is therefore going to be a lot smaller than a one carat diamond. It is an insulator, whereas diamond is a good conductor of heat. These are all things that can easily be determined with the right equipment. More recently, another synthetic mineral that is exceedingly rare in nature, moissanite (silicon carbide, SiC) has entered the gemstone market, and with a much greater hardness (9.5) it is an even better diamond-imitator. However, like CZ, it has a range of other properties that distinguish it from the real thing.

Some minerals are regarded as semi-precious in terms of their colour and texture rather than colour, clarity or play of light. These are often obtained in bulk so that large flat pieces may be cut and polished as the starting-point for a variety of uses. Malachite ($Cu_2CO_3(OH)_2$), turquoise ($CuAl_6(PO_4)_4(OH)_8 \cdot 4H_2O$), lapis-lazuli (a mixture of lazurite, $(Na,Ca)_8[(S,Cl,SO_4,OH)_2|(Al_6Si_6O_{24})]$ and other minerals) and rhodochrosite ($MnCO_3$) are all examples. Malachite and turquoise are secondary copper minerals found in the oxidation-zones of copper ore deposits; lazurite is a rock-forming mineral; and rhodochrosite is a primary mineral found in hydrothermal deposits. A wide range of other minerals are likewise of value when occurring in a sufficiently massive form. Malachite and rhodochrosite are prized for their light- and dark-coloured concentric banding. Their relative softness means that they can easily be carved or turned; thus there are almost limitless possibilities in their application, with good-quality material having a unit value far greater than their worth as ores or industrial minerals. Such materials are often used for interior decoration, as are large amethyst **geodes**, worked in the lava-flows that contain them in many parts of the world. Geodes a metre or more in height – known as 'cathedrals' – are often the basis of expensive internal lighting features, as opposed to simply being mineral specimens.

7 Minerals and the environment

7.1 Overview

It is an ironic thing that minerals, omnipresent in the natural environment, both cause and cure when it comes to pollution. According to the Oxford English Dictionary, the definition of pollution is: 'the presence in or introduction into the environment of a substance which has harmful or poisonous effects'.

Harmful or poisonous effects clearly refer to effects upon animal and/or plant life, and affecting both wild and cultivated species and humans, depending on the physical and chemical properties of any one occurrence of pollutants. Substances are widely variable in their toxicity: selenium, for example, is in tiny amounts an essential dietary trace-element for mammals; in moderate amounts it causes severe toxicity. Hydrogen sulphide, with its rotten egg odour, can be detected by most people at just 0.0047 ppm in air. At concentrations of over 500 ppm it is seriously toxic and potentially lethal. Carbon dioxide is essential to photosynthetic plant life (and therefore ourselves); however, it has other properties which, in contrast, make it dangerous. As a strong greenhouse gas, increasing atmospheric carbon dioxide levels from a pre-industrial 280 ppm to 400 ppm or more make it a pollutant because of the impacts of rapid climate change; at much higher levels it becomes an asphyxiant, as tragically evidenced in 1986 at Lake Nyos, in Cameroon. Here, the magma underlying an old volcanic crater-lake gives off carbon dioxide, with which the lake water becomes saturated. At depth, the confining pressure of the water-column keeps the gas in solution, but any triggering mechanism that forces a lot of that deep water upwards into lower pressure zones can cause it to explosively degas. In 1986 it was thought that landslides were the trigger: the consequence was that Lake Nyos suddenly emitted a large cloud of carbon dioxide, which, due to its relative density, rolled along the ground immediately after the eruption, displacing the air as it did so. Over 1700 people and 3500 livestock died from asphyxiation in nearby communities.

These examples, all involving everyday substances, demonstrate that it is the concentrations of different substances with different properties that determine whether they are harmless and/or beneficial or pollutant in nature. So where do minerals come into the equation? In two ways: firstly, in some cases, as a cause, and secondly, in fewer but potentially useful cases, as a cure.

7.2 Mineral-related pollution: causes

In general, natural mineral occurrences rarely constitute pollution. Large-scale occurrences of common minerals are typically problem-free. Take quartz, for example. It is generally harmless with a few exceptions, such as burial of animals and plants in severe sandstorms. Chemically inert to all intents and purposes, it harmlessly makes up much of the millions of square miles of beaches around the world. The same point applies to other common rock-forming minerals such as feldspar and mica.

Erosion and/or weathering of natural occurrences of more toxic minerals rarely cause major problems. Take galena, for example: a vein of this lead sulphide, cropping out in a sea-cliff, may be affected by rockfalls and other forms of erosion, but the quantities of mineral released to the environment are, on the overall scale of things, relatively small. Where groundwaters do contain elevated metals, for example over some disseminated, porphyry-style copper deposits, one quite often finds that specialized floral communities have developed through time. A famous example is at Coed y Brenin in North Wales, where a porphyry-type copper deposit of lower Palaeozoic age has a late Quaternary glacial erosion-surface. Active chemical weathering since that erosion period has taken place over a large area of bedrock containing dissipated chalcopyrite and other copper-bearing sulphides; the copper became mobilized into groundwaters. The copper-bearing solutions were subsequently reduced and deposited as the native metal and various salts in surrounding peat-bogs. In some of the bogs, the peat was actually dug profitably as copper-ore in the nineteenth century. The flora of these boggy areas is remarkable; it includes a number of plants, such as pink thrift and white bladder-campion, which are normally found on the coast. Here they are thriving, and clearly have been for a long time, at the expense of other inland plants whose intolerance of elevated metal concentrations prevented them from colonizing the area widely (Fig. 7.1).

Contrast such slow processes with mining, using metals as a commonplace example. An ore-body may take between a few years to a few decades to remove and process, depending upon its size and the type of mineralization. There will be three key product-categories that result from the operation. Firstly, the commodities worked, which are largely taken off-site to be refined and marketed. Secondly, the **development rock** – that rock which is removed during the construction of shafts and levels through which to access ore-bodies. This may or may not be mineralized, depending on the geology of the mine, and it will be dumped, generally at surface, in fragments ranging from a metre or two down to dust-sized. Thirdly, the tailings – fine-grained waste that is the residue from the operation of crushing, grinding and extracting the commodities from the ore. Typically in the form of a watery slurry, tailings tend to be dumped into lagoons built for containment purposes (Fig. 7.2).

The very process of mining, crushing and grinding mineralized rock is what makes pollution more of a risk.

Figure 7.1 A copper-rich peat-bog in North Wales
The 'Turf Copper' mine in Coed y Brenin forest, North Wales. Although the copper-impregnated peat was dug away many years ago, levels of the element are still high enough to provide an ecological niche for a metal-tolerant flora. Pink sea-thrift is accompanied by the white sea-campion and the smaller white vernal sandwort, all thriving miles away from the more usual habitat by the coast.

That is because the surface areas of any reactive minerals like sulphides are vastly increased by the processing: something that can be visualized with a simple example. Imagine a cube of a reactive substance – iron, for example, 10 centimetres on edge. To calculate its surface area, we first obtain the area of one face: 10 cm x 10 cm = 100 cm². As a cube has six identical sides, we multiply the area of one face by six, thus getting 600 cm² of iron waiting to react with moist air and go rusty. But what happens if we divide the original cube up into 1 cm cubes? The cube's volume is 10 x 10 x 10 = 1000 cm³, so once divided, we have a thousand 1 cm cubes. Each will have a surface area of 1 x 1 x 6 = 6 cm². The total surface area of all the cubes will be 6 x 1000 = 6000 cm². By dividing the original cube up into all these little ones, we have increased the surface area available to react by a whole order of magnitude. Divide those centimetre cubes into millimetre cubes and we now have 60,000 cm² of surface available to react, and so on.

Now apply that concept to mine waste, full of pyrite – which it frequently is – plus lesser amounts of other metal sulphides. Pyrite decays when exposed to air and moisture, liberating sulphuric acid in the process, which in turn can

Figure 7.2 Tailings
The tailings dump of an old lead-mine in Mid-Wales: an extensive expanse of fine (2–3 mm down to dust-sized) crushed and powdered mineralized rock, the waste residue from ore-dressing. Inefficiencies in the process mean that the tailings still contain considerable amounts of lead, zinc and other metals and the toxicity is so high that virtually nothing will grow over most of the area.

Figure 7.3 Ochre
Acid mine drainage can be a major environmental problem. The water that normally flows out of this adit at Cwmrheidol mine in Central Wales (photo taken during a prolonged drought) typically has a pH of just over 2 and is rich in dissolved zinc, cadmium and other undesirable substances.

attack other sulphides and accompanying minerals, so that the result is acidic water laced with a cocktail of whatever metals were present in the ore-deposit. That is why pollution can be a major problem around mines, especially older ones, which may have been in operation before any environmental regulations were in place. Metalliferous leachates from tailings are one problem; another is acid mine drainage, where due to sulphide decay taking place within a mine, groundwaters leaving the mine are loaded with enhanced metals.

Pollutants from such situations vary greatly: at copper mines, for example, arsenic may be among the pollutants because the two elements often occur together in ore mineral assemblages. At zinc mines, cadmium pollution can be a problem because the metal occurs most commonly as an impurity in the common zinc sulphide, sphalerite. Very acidic minewaters can even leach elements from the deposit's host-rocks, bringing things like aluminium (derived from alumino-silicates) into solution, driving up the toxicity again. Such chemical cocktails can cause severe ecological damage to river-systems, in the worst cases leaving them devoid of life (Fig. 7.3).

Particulate pollution from tailings can sometimes cause problems, especially in the vicinity of older mining areas. Very fine-grained, part-oxidized sulphides can be harmful if inhaled or ingested. In the 1970s, a problem was discovered at an old lead-silver mine at Cwmsymlog in Central Wales, where during a very dry summer local children had been playing among heaps of very fine tailings – so-called slimes. The dust inhaled by them had raised their blood lead concentrations worryingly. The tailings area was subsequently subject to remedial works, the first of a series of such works in the district. In 2012, prolonged heavy rains and major flooding affected a nearby area where there were further mines. Fields along the affected floodplains were subsequently cut for silage, and when that was fed to cattle a number of them sickened, and some of those died in turn from lead poisoning; the floodwaters had eroded and then spread finely-divided tailings over the riverside fields, contaminating the grass with lead and other metals.

Another problem that can affect old mining districts is the collapse of old workings, impeding the easy flow of groundwaters, which may build up until the head of pressure reached overcomes the temporary dam created by the fall, resulting in a blowout. If these waters are badly polluted, such an incident can have disastrous consequences. To cite a third Central Wales example, the Cwmrheidol mine consists of two adits driven into a steep hillside to access a sulphide-bearing vein system. The veins carry much marcasite, among the other base-metal sulphides, which is an even less stable iron sulphide than pyrite. Abandoned long ago, by the mid twentieth century the entrance to the lower adit had collapsed and even its exact location was uncertain. It was inadvertently rediscovered in the late 1960s by someone using an excavator to create a flat area upon which to situate a drilling rig; an enormous volume of highly acidic, metal-laden water accompanied by huge quantities of iron **ochre** burst forth. Fish life in the nearby river, Afon Rheidol, was wiped out instantly, and the inshore waters off Aberystwyth, ten miles downstream where the river enters the sea, were coloured bright orange by the incident. In the early 1990s, following the discovery

that a new blockage had closed the adit again, contractors were brought in to carefully discharge the water in a controlled manner, before removing the dam; this was considered preferable to a further sudden blowout (Fig. 7.4).

Figure 7.4 Cwmrheidol mine
Emergency works to prevent a sudden blowout of highly polluted, acidic water from Cwmrheidol mine, in the early 1990s. The water and ochre were pumped out round the clock over several days and the collapse damming the tunnel was removed and the tunnel lagging replaced.

Old smelters also cause problems, as in pre-regulatory times the by-products were simply released to the atmosphere. County-scale geochemistry maps often reveal areas of such past activity, especially in areas where the British Geological Survey have completed their Geochemical Baseline Survey of the Environment (G-BASE). Their geochemical atlas for South Wales revealed an interesting arsenic anomaly extending to the north-east of the Swansea district; this was briefly perplexing in geological terms, until it was realized that a large amount of the ore mined in Devon and Cornwall – where chalcopyrite frequently occurs in association with arsenopyrite – was historically smelted in the vicinity of Swansea. Arsenic is a volatile element and the prevailing wind is from the south-west: QED.

Gold extraction poses its own problems. Gold that is too fine-grained to be extracted by gravity alone (washing the powdered gold-bearing quartz with water) has for centuries been extracted by grinding the ore with mercury. The two metals readily amalgamate, and when the amalgam is saturated with gold it stiffens noticeably. It is then scraped up, and heating it drives off the mercury as vapour, leaving the gold behind. This is where the problems occur. Although it is possible to recover the gaseous mercury using condenser technology, in many artisanal mining districts around the world there are few such precautions taken, and the imperfect processing exposes people to the mercury vapour – which is highly toxic and has in many cases proved fatal. Such problems have made the news frequently in recent years in remote parts of South America. Gold also dissolves in a solution of sodium cyanide; cyanidation has replaced amalgamation in modern gold extraction, but again this is a toxic reagent whose effects both on health and the environment can be devastating.

Misuse of some minerals has led to major problems, of which perhaps the most notorious one concerns asbestos. This is a group name for a series of minerals, belonging mostly to the amphibole group, which tend to crystallize in a very finely fibrous form. Of these minerals, crocidolite, which is the name for the fibrous form of riebeckite $(Na_2Fe^{2+}_3Fe^{3+}_2Si_8O_{22}(OH)_2)$, also known as blue asbestos, is regarded as the most hazardous. The heat-retarding properties of the asbestos group have been well-known for a long time and by the mid-twentieth century uses included fire retardant coatings, fireplace cement, brake-linings, resistant gaskets, pipe insulation, safety curtains and many others, despite the fact that its toxicity had been noted as far back as Roman times. Asbestos-group minerals cause a range of lung problems and are highly carcinogenic; since the detailed epidemiology of asbestos-related diseases became understood, usage has dwindled significantly, and whenever blue asbestos is discovered in older buildings in the UK, it is removed by specialist contractors.

Finally, the problems of safe storage of radioactive mineral specimens because of the hazard of radon gas have a parallel in the wider world. Some rock-types, especially granites and organic-rich black mudstones, have enhanced radioactivity. This is because of the presence of small amounts of naturally-occurring radioactive elements such as uranium, which may be present within discrete mineral grains such as uraninite (UO_2), especially in granites. In black shales, uranium is often enriched in their organic component if the element was present in solution within the depositional environment. Natural radioactive decay of uranium produces radon as a daughter-product; the radon itself goes on to decay via a chain of other radioactive elements to a stable isotope of lead. Another widespread (in small amounts) radioactive element, thorium, likewise includes an isotope of radon in its decay-chain. The solid elements along these decay-chains are, unless mobilized – for example if the wind blows dust around – relatively harmless because, away from rich sources of uranium mineralization, the amounts are so tiny. Radon, however, is more mobile and dangerous amounts of it can accumulate in unventilated areas, where there is a risk of people breathing it in. Such areas might be blind tunnels in mines that have no through-flow of air, or basement rooms in buildings. In both cases the local geology is of critical importance – the rocks near and at surface need to be enriched in uranium or thorium. Such properties are routinely recorded during modern geological mapping, so that areas with potential radon hazards are well defined in the UK, and appropriate measures can be taken to eliminate the risk, such as forced ventilation. However, this does not extend to old mines a century or more old – something which experienced mine-explorers will take into account.

7.3 Mineral-related pollution: cures

In many cases, the simplest way to prevent mineral-related pollution is to immobilize the pollutant elements at source, to prevent them escaping into the wider environment in mobile forms such as aqueous solutions. A good example is mine-tailings, which may be sealed into heavy-duty plastic sheets – geomembranes – which radically reduces the degree of their interaction with air and water. Such sealing is a common method at old metal-mines where the oxidation and dissolution of metallic sulphides and sulpharsenides has led to pollution problems in adjacent watercourses. Burial of sealed-in tailings also stops the wind from mobilizing them in dust form.

At some old mines, the technology of yesteryear was inadequate in recovering all of the ores, especially from fine-grained **polymetallic** ore-bodies. This poor recovery may have been down to the fine grain-size itself, or the presence of one or more minerals, for example with very similar specific gravities to the sought-after mineral, which would have hindered the recovery processes. Whatever the cause – and there are many – the result would be mine-tailings that still contained considerable quantities of valuable elements, and with the application of modern technology, it has often been possible to rework them in an economically viable way, leaving behind a significantly less metalliferous waste and recovering large amounts of useful commodities in the process. Today, the use of modern technology in artisanal mining fields is to be encouraged, because it is the best counter to the risks of pollution and poisoning by mercury, cyanides and other toxic separation-reagents.

It is one thing dealing with a source of pollution when that source can easily be got-at; it is quite another thing to deal with pollution when its source is deep underground, and either difficult or impossible of access. The aforementioned Cwmrheidol mine in Central Wales is a classic example. A large quantity of water makes its way into the mine along near-surface workings where the mineral lodes crop out on the hilltop above. The water picks up metals as it makes its way through the old workings and then escapes back to surface via the main drainage-adit and into the Afon Rheidol, an important salmon and sea-trout river. Early attempts to strip the mine discharge of some of its worst properties involved passing the discharge through a filter-bed of limestone chunks. However, this was not a long-term success: the sulphuric acid that gave the discharge its low pH reacted with the limestone to form gypsum, which being only slightly soluble, built up as a protective coating on the surfaces of the limestone.

In 1993, the situation was compounded when a collapse in the uppermost old workings of the mine captured an entire stream, massively increasing, albeit diluting, the discharge. However, this temporary problem was simply overcome by bridging the stream over the collapse. The real challenge remained in dealing with the normal discharge. Recent experiments, however, have found that by putting the water through a filter-bed containing compost, limestone chips and cockle-shells – all readily available products – a bacterial ecosystem develops that strips out over 95% of the **metal loading** (mostly zinc, cadmium and lead). If this can be adapted on a much larger scale at Cwmrheidol and other problematic metal-mines, it will represent a major advance in pollution-control. Furthermore, it may be possible to recover the metals from the filter-bed material whenever that is replaced – at Cwmrheidol that would amount to tens of tons of zinc per year alone. A much simpler use of chemistry used to be deployed to great advantage at the copper mines at Parys Mountain on Anglesey: the copper-rich and thereby toxic mine discharge was run through tanks into which any old scrap iron that was available was placed. By **double decomposition**, the iron was totally replaced by metallic copper, which was periodically scraped out of the tanks as a brown sludge and refined at profit.

Rendering **mobile** pollutants into **inert** compounds is another developing branch of pollution control technology. Some interesting research has been undertaken in recent years with mobile, and therefore **bioavailable**, lead in soils. Treating the ground with various additions of phosphate leads to the development of much more stable lead-bearing phases such as the mineral pyromorphite, $Pb_5(PO_4)_3Cl$, which is actually insoluble in the intestinal tracts of animals: a clear case of a mineral being a potential cure for pollution (Fig. 7.5).

These examples all point to two basic principles with regard to mineral-based pollution. Firstly you try to isolate it at source, but if that proves impossible, then you need to find a way of detoxifying it before it ends up where it becomes a real problem.

Figure 7.5 Pyromorphite
The lead chlorophosphate mineral, pyromorphite, is the most stable lead-bearing compound in nature. Research has been undertaken involving its artificial creation in soils contaminated with mobile lead pollution by treating them with various phosphate compounds, in an attempt to render the lead immobile. These greenish crystals on quartz come from the oxidation-zone of a lead-bearing vein deposit in Central Wales.

7.4 Mineral-related environmental benefits

Although mineral-related pollution can be a serious problem where abnormally high concentrations of certain elements occur, it is important to note that without background levels of minerals, the planet would look very different indeed, for the simple reason that soils would not be as they are. Soils are complex growing mediums consisting of rock and mineral particles from clay-sized up to boulders, mixed with organic matter and held together by mineral cements and the roots of plants and, when undisturbed, containing networks of pore-spaces through which water and dissolved minerals can move. In this environment there exists a great diversity of microbes that recycle organic matter, acting as agents for nutrient storage and release during their life-cycles.

Plants – including those we grow for food – require a number of essential chemical nutrients to be healthy and thus crop well. In the absence of man-made fertilizers, carbon (as carbon dioxide), oxygen and nitrogen are obtained directly or indirectly from the atmosphere; the remainder, phosphorus, potassium, calcium, sulphur, magnesium and trace elements such as boron, chlorine, manganese, iron, zinc, copper and molybdenum, are derived from weathering of the minerals present in the soil. In non-glaciated or non-alluvial areas (such as on the slopes of an extinct tropical volcano), soil chemistry will directly reflect that of the underlying bedrock, but in formerly glaciated areas and along river floodplains, a wider mixture of rock-types (and thereby mineral species) may be available, in a whole range of grain-sizes, to weathering-agents, often resulting in soils of better growing quality because the full range of nutrients is present in quantity.

Clay-minerals, which are sheet-silicates such as montmorillonite ($(Na,Ca)_{0.33}(Al,Mg)_2(Si_4O_{10})(OH)_2{\cdot}nH_2O$), to name but one of many examples, are especially important in soils. Formed by the chemical weathering of other silicate minerals such as feldspars or pyroxenes, their microscopic crystals attract and hold water and they act as reservoirs for plant nutrients. This vital role works because clay-mineral particles carry unbalanced electrical charges at their surfaces so that some attract cations (such as Ca^{2+}) and some attract anions (such as $HPO4^{2-}$), which build up on them, rather than being entirely lost via soil drainage. In soils that are very sandy this loss is a major problem: they dry out too quickly and they are nutrient-poor. Remember the cube of iron being cut into smaller particles and increasing its surface area in the section on weathering? The same thing applies here; soil rich in tiny particles of clay (which are defined as being 0.002 mm across or less) has a vastly greater internal surface area available to hold onto nutrients than soil that is mostly sand (the grain-size of which is defined as being between 0.063 mm and 2 mm).

In the tropics, things are rather different because the weathering does not just stop when clay-minerals are formed; they themselves break down, leaving behind oxides of aluminium and iron such as gibbsite $(Al(OH)_3)$ and goethite $(Fe^{3+}O(OH))$ respectively – which is why tropical soils are very thick (due to the great depth to which the weathering penetrates) and typically reddish-brown. The oxides are very poor at holding onto soil nutrients compared to clay-minerals. In tropical rain-forests, that nutrient-reservoir role is almost entirely played by humus, which is swiftly removed by rainfall when a forest is felled, leaving

a landscape that, in terms of growing food-crops, is next to useless, as anyone trying to grow a plant in a pot of powdered and blended gibbsite and goethite will discover.

In more temperate climates where the weathering regime is generally less severe, clays tend to be preserved. However, the quantity of clay-minerals present is not only important in determining the soil's nutrient-holding capability; it also determines the nature of the soil's structure. Soils with too much clay-mineral content are often referred to as 'heavy' and they tend to become waterlogged too easily in wet weather. To the gardener, an ideal composition, referred to as loam, is where the mineral particles in the soil, of sand (2–0.063 mm), silt (0.063–0.002 mm) and clay-size (<0.002 mm) are present in equal amounts, a rare thing indeed in nature! (See Fig. 7.6.)

Even where fertile, nutrient-rich soils with the optimum clay-mineral content are present, such as in the Great Plains of the American Midwest, monocultural agriculture tends to deplete certain nutrients preferentially, depending on the type of plant being grown, and such practices can only be maintained at a reasonable yield by artificially replenishing such nutrients by the addition of chemical fertilizers. Given that such fertilizers are based on feedstocks originating from finite mineral resources, it is unlikely that monoculture has a long-term future in many parts of the world: one of a number of potential problems stemming from the way we use our mineral and other natural resources at present. A full understanding of the essential roles played by minerals in soils, and how minerals and plants interact, is going to be essential in tackling such issues effectively.

Figure 7.6 Good soil
A temperate climate, good soil mineralogy and regular additions of composted organic matter, containing essential and trace-elements, are all helpful when it comes to growing edible crops.

Epilogue

In this book, I have tried to capture the true diversity that exists in the field of mineralogy. Minerals and their properties are of relevance to almost every aspect of our daily lives; whether you go on to undertake PhD research, start a career in economic or environmental geology, become a mineral collector or dealer, or decide to go into something completely different like agriculture, you will never be far away from some aspect or another of the science.

Over the years, mineralogy has given me much pleasure, fascination, and occasionally excitement, as in the case of finding the gold nugget described at the start of the book. That was the product of a long chain of events starting with my childhood, where an uncle worked in the fluorspar mines in the English Pennines and on visits I would be taken collecting. 'A' levels were followed by undergraduate years at Aberystwyth, whose hinterland, with its many old metal mines, soon proved to be a happy hunting-ground, and became the subject of my postgraduate research. Through this interest in mineral deposits and a desire to see more of them, I volunteered for the British Geological Survey's Mineral Reconnaisance Programme, where I wandered the wilds of Scotland, learned about gold deposits and, importantly, how gold particles of all sizes behave when they get into rivers. Without that chain of events taking place, someone else instead would, without doubt, have found that Welsh nugget!

When writing this book with, as is always the case, a word-limit, it was difficult at times to decide what to include and what to leave out. Somebody else with the same task might well differ in their selection, which will always be based to some extent on personal experience. However, I hope that reading these introductory chapters has provided a good, broad grounding and, perhaps, inspired you to look into the subject in greater depth.

Glossary

A

accessory minerals [35]: the minor minerals (such as fluorite or apatite) that occur in igneous rocks that are predominantly made up of feldspar, quartz, pyroxenes and so on.

accretion [27]: in planet formation, the collection of material through gravity that over time builds up planetary mass.

achondrites [29]: stony meteorites of an evolved nature, consisting of igneous rocks such as basalts. Thought to have originated via large collisions between planetary bodies where some differentiation had already taken place. The rare Lunar and Martian meteorites fall into this category.

acicular [10]: term to describe crystals that have a thin, needle-like habit.

acidic rocks [33]: igneous rocks containing a high silica content (greater than 63% SiO2) such as rhyolite and granite.

adamantine [21]: term to describe minerals with a brilliant, sparkly lustre.

algal bloom [50]: an algal bloom may occur when water is overloaded with nutrients such as nitrates or phosphates. The algae multiply dramatically, typically use up the nutrients, and then have a mass die-off: their decomposition may then cause significant de-oxygenation of the water.

alkali metal [3]: any of the group 1 elements – lithium, sodium, potassium, rubidium, caesium: all highly reactive monovalent metals.

alkaline earth metal [3]: any of the group 2 elements – beryllium, magnesium, calcium, strontium, barium: all highly reactive bivalent metals.

allogenic [36]: a mineral grain occurring in a sedimentary rock that has been transported from elsewhere to the place of deposition e.g. alluvial gold occurring in a sandstone.

alpine fissure-type mineralization [49]: a highly localized and variable form of hydrothermal vein mineralization that typically develops in brittle fractures in competent lithologies occurring in an otherwise ductile regime, for example in a dolerite dyke cutting mudstone that then undergoes compressive deformation leading to the formation of slate.

amorphous [1]: any non-crystalline mineral that will not yield a useful X-ray diffraction pattern.

amphiboles [27]: a group of silicate minerals with the generic formula: $XY_2Z_5((Si,Al,Ti)_8O_{22})(OH,F,Cl,O)_2$ in which X is sodium, potassium or calcium, Y is sodium, magnesium, ferrous iron and calcium (less commonly lithium or manganese) and Z is sodium, magnesium, ferrous or ferric iron, aluminium or chromium (less commonly lithium, manganese, zinc, cobalt, nickel, vanadium, titanium or zirconium).

analyser [75]: the upper of the two polarizers in a polarizing optical microscope: it can be inserted and removed by the operator. Insertion means that the operator is looking at the thin section through cross-polarized light and the interference-colours and extinction properties of the minerals may be studied.

ångström [65]: the ångström is a unit of length equal to 10^{-10} m (or 0.1 nm). This ultra-microscopic size range is used for describing the sizes of atoms, molecules etc.

anhedral [10]: of crystals: a crystalline mass without any regular shape.

anisotropic [76]: in petrography, refers to minerals in thin or polished sections whose optical properties (e.g. colour, brightness) are directionally dependent as opposed to identical in all orientations (isotropic minerals).

anisotropy [76]: term to describe how an anisotropic mineral appears under the polarizing microscope: it is generally used in a subjective manner, e.g. weak, strong and so on.

anorthosite [42]: an igneous rock made up almost entirely (>90%) of plagioclase feldspars.

anthropogenic [1]: anything occurring as a result of human activity.

aphanitic [32]: any igneous rock that is too fine-grained for the constituent crystals to be detected by the naked eye.

arsenopyritization [45]: a form of wall-rock alteration involving replacement of original minerals by arsenopyrite. Commonly associated with many gold deposits.

asphyxiant [97]: a non-toxic or minimally toxic gas that either reduces or displaces the normal oxygen concentration in air, the breathing in of which can then lead to death by asphyxiation (suffocation).

associations [65]: groups of minerals that are frequently found occurring together, such as galena-sphalerite-fluorite-barite.

asterism [21]: an optical phenomenon displayed by some gemstones such as sapphires, in which a bright star-shape is visible on correctly-cut cabochon faces. The phenomenon is caused by light being reflected from myriad needle-like orientated inclusions of rutile occurring within the stone.

Asteroid Belt [27]: the region of the Solar System situated approximately halfway between the orbits of the planets Mars and Jupiter, and occupied by large numbers of mostly irregular rocky bodies – the asteroids, varying from small chunks of rock up to much larger bodies such as Ceres, which is *c*.950 km in diameter.

atom [1]: a basic unit of matter that consists of a dense central nucleus, made out of positively-charged protons and non-charged neutrons, which is surrounded by a cloud of negatively charged electrons, the positive and negative electrical charges binding the whole thing together.

atomic number [2]: of an element, the number of protons contained in the nucleus of any of its atoms: e.g. gold has an atomic number of 79, meaning there are 79 protons in the nucleus of any atom of gold.

authigenic [36]: a mineral occurring in a sedimentary rock that has crystallized *in situ*, e.g. calcite in a limestone.

aventurescence [21]: a light-scattering phenomenon caused by inclusions of one mineral within another: for example, aventurine is quartz crowded with preferentially-oriented small plates of the green, chromium-bearing mica fuchsite, which make the quartz appear to be full of green glitter.

axial cross [11]: the central meeting point of the axes of any crystal.

B

backscattered electrons [79]: in scanning electron microscopy, high-energy electrons originating from the SEM beam itself that have been reflected from the sample via interactions with the atoms making up its surface.

banded-iron formation (BIF) [50]: a sedimentary rock that typically consists of alternating thin layers of haematite, magnetite and silica, associated with mudstones and cherts of marine origin. An important source of iron, they are generally restricted in occurrence to Precambrian strata, with the most extensive occurring in sequences dating from the early Proterozoic, between about 2.5 and 1.8 billion years ago.

basal cleavage [ ]: a cleavage that occurs parallel to the base of a crystal.

basement [41]: generic term referring to ancient, typically crystalline, rocks that occur at depth beneath the geology exposed at surface.

base-metals [4]: informal term covering any metal that readily oxidizes at Earth's surface and reacts variably with diluted hydrochloric acid to liberate hydrogen. Iron, manganese, copper, nickel, zinc and lead are all considered to be base-metals.

basic rocks [34]: also referred to as *mafic rocks* for their magnesium and iron content, basic rocks are relatively silica-poor (45–52% SiO_2). Examples include basalt and gabbro.

bioavailable [102]: any element or compound present in the environment, that may be taken up by living organisms.

bipyramid [8]: a crystal whose vertical axis is terminated at either end by pyramid faces. Frequently found in the hexagonal and tetragonal crystal systems.

bireflectance [76]: of anisotropic minerals viewed in polished section under plane-polarized light: changes in reflectivity observed as the microscope stage is rotated and described subjectively e.g. weak or strong.

birefringence [75]: many minerals polarize incoming light as a result of their structural chemistry. Often, the light is split and polarized into two different planes of vibration – the double refraction, or birefringence, of Iceland Spar calcite being the classic example.

black sand [53]: various concentrated oxides of iron, titanium and so on that have accumulated in placer deposits: often found accompanying alluvial occurrences of gold, cassiterite, etc.

black smokers [46]: submarine hydrothermal vents through which hot, metal-charged brines are expelled. The changes in temperature, pressure and chemistry when they reach open water cause sulphides to precipitate out as fine particles, hence the reference to smoke: these particles may in time accumulate as chemical sediments to create significant ore deposits.

bladed [10]: crystals whose shape is similar to a knife-blade.

body-centred cubic unit cell [7]: a cubic unit cell in which eight identical particles are situated at its corners and a ninth identical particle is situated right at the centre of the body.

botryoid [15]: a grape-like body, internally consisting of often layered, fibrous crystals that have grown out from a common centre.

boudinage [82]: the process by which a rigid body occurring in a ductile matrix deforms when the compressive stress is at a steep angle to its surface: it responds by extension and the end result is typically a series of pod-like bodies (boudins) separated by much thinner zones (the boudin necks) where there may also be brittle fracturing. The end result looks rather like a string of sausages.

breccia [45]: shattered wall-rock cemented together by minerals: often found along faults and in mineral-lodes.

bulk dealers [68]: mineral dealers who import certain popular minerals such as amethyst in multi-tonne shipments, often selling the material on in wholesale batches.

bund [59]: a bund is an elongate earthen mound erected usually as a barrier, e.g. in working quarries, to prevent access to hazardous rock-faces.

C

calcium chloride [62]: a highly hygroscopic compound that may be used in the storage of specimens of other, less hygroscopic, minerals in order to absorb any moisture in the vicinity, thus helping preserve them.

carat [32]: unit of gemstone weight measurement: 1 carat = 200 mg or 0.2 g.

carbonatites [35]: unusual igneous rocks made up of 50% or more carbonate minerals.

carbonatization [45]: a common form of wall-rock alteration around ore deposits in which minerals are replaced by carbonates such as calcite and/or dolomite.

carbon-sink [36]: any natural or artificial reservoir that accumulates and stores carbon-bearing chemical compounds for an indefinite period of time: an excellent long-term example being limestones, which store carbon typically for millions of years.

cataclasis [41]: the pervasive internal shattering of rocks that can occur along fault-zones: a form of dynamic metamorphism.

cements [38]: typically used in sedimentary petrology to describe the minerals binding together the individual grains making up a sedimentary rock. Can also be applied to mineralized breccias, where various minerals are described as cementing together the clasts of brecciated rock.

chatoyance [21]: an optical phenomenon displayed by minerals which either occur in fibrous bands or contain fibrous bands of other minerals. Tiger's eye is a good example, consisting of quartz containing bands of fibrous silicates of the amphibole group. The fibres scatter the light so that the bands have a luminous appearance which seems to move as a specimen is turned in the hand.

chemical compound [1]: a pure chemical substance that consists of two or more different chemical elements in combination.

chemical feedstocks [87]: minerals that are used as raw ingredients for the production of useful chemical compounds: e.g. fluorite is a raw ingredient for the process in which hydrofluoric acid is manufactured.

chloritization [45]: a frequently-seen form of wall-rock alteration around mineral deposits in which minerals of the chlorite group variably replace the wall-rocks.

chondrites [29]: stony meteorites that have not been modified due to melting or differentiation of a parent protoplanetary body. They are thought to have formed when various types of dust and small grains that were present in the early Solar System accreted to form primitive asteroids. An important feature is the presence of chondrules, which are round grains formed by distinct minerals including various silicates, and are believed to be among the oldest discrete bodies in the Solar System.

clastic [35]: pertaining to any sedimentary rock made up of rocky detritus – clasts – which may be irregular or rounded and may be from dust up to boulder-size, depending on the rock-type.

cleavage [5]: in mineralogy, planes of weakness due to the structure of the crystal lattice and the relative strengths of the various chemical bonds holding it together. Can be perfect in some minerals, so that crystals will cleave into flat sheets, and non-existent in others, with a whole range of terms (imperfect, indistinct, poor, etc.) in between.

cockscomb [15]: aggregates of platy or tabular crystals arranged on edge in a criss-cross fashion.

columnar [10]: long, stout prismatic crystals aggregated together.

conchoidal [18]: the fracture pattern displayed by some minerals (such as quartz): the fractures curve like a clam-shell with repeated ridges of a similar nature to those seen in broken glass.

concretion [38]: a nodular mass, rounded to irregular and of variable size, of a typically fine-grained mineral (or minerals) occurring embedded in a sedimentary rock, such as clay-ironstone (siderite plus clay minerals) nodules commonly found in coal-bearing strata.

conglomerates [35]: clastic sedimentary rocks consisting of large (2 mm or often much bigger) pebbles cemented together, often with a finer-grained (e.g. sandy) matrix.

connate brine [47]: connate waters are those that are trapped within sedimentary rocks during their deposition and burial. Occupying the interstitial pores between grains of sediment, they are often liberated at a later stage by, for example, low-grade regional metamorphism, making them a potential source for mineralizing hydrothermal fluids.

contact metamorphism [39]: the weak to intense changes in mineralogy brought about by the thermal cooking (or baking) of rocks intruded by a magma. The degree of metamorphism depends both on the characteristics of the intrusion (size, temperature of magma, etc.) and on the types of wall-rocks that are being intruded.

continental crust [34]: that part of Earth's crust that makes up the continents and their shelves, comprising metamorphic, igneous and sedimentary rocks with an overall bulk composition close to that of granite and a thickness ranging from about 25 to 70 km.

core [30]: the metallic inner zone of Earth, some 3486 kilometres in thickness, and divided into a solid inner core (1220 km) surrounded by a liquid outer core (2266 km), as deduced by their different interactions with seismic waves.

covalent bond [5]: strong chemical bonds formed between non-metals whose atoms co-operate by sharing outer-shell electrons to obtain a full outer shell, rather than one atom giving an electron and the other receiving it, as in ionic bonding. Covalent bonds can form between atoms of the same element to create molecules, such as H_2 or Cl_2, or between different elements to form molecular ions such as SO_4^{2-} or PO_4^{3-}.

crust [25]: the relatively thin layer of almost entirely solid rock that forms the outermost part of Earth.

crystal lattice [5]: a three-dimensional structure of packed-together atoms and/or molecules whose geometrical characteristics are determined by the sizes of the various atomic and molecular ions involved in different compounds.

crystallographic axes [7]: the axes of rotational symmetry, the ratios of which describe the shape of a crystal in terms of length.

crystal system [7]: crystals are divided into six crystal systems defined by their morphology, something which is controlled by the ratios of their crystallographic axes, and which in turn defines their planar and rotational symmetry properties. Each mineral species crystallizes in one of these six systems: the shape any crystal takes in its outward appearance is known as its habit.

cubic or isometric system [7]: a crystal system in which all three crystallographic axes are of equal length, situated at right angles to one other. Thus equidimensional (or equant) crystals such as cubes or octahedra are formed – these being two of the commonest habits in the cubic system. All crystals belonging to this system have four 3-fold axes of rotational symmetry, each running diagonally from corner to corner through the centre of the cube. They may also have up to three 4-fold axes of rotational symmetry connecting the centres of each of the three opposite pairs of faces.

cubic zirconia [95]: synthetically manufactured zirconium dioxide, a material which has some similar properties to diamond, and which is used as a cheaper substitute for that mineral in some jewellery.

cumulate [35]: an igneous intrusion in which differentiation of heavy and light minerals (such as chromite and anorthite respectively) occurred as they crystallized from the melt and either sank down or floated up through it to accumulate in a layered fashion.

cut-off grade [87]: that grade below which mineralization is not economically worth mining as an ore. Can be affected both by geology and by external factors such as fluctuating global metal prices.

D

deads [60]: in mines, especially underground: walled-up, often timber-supported stacked waste rock. In old mines, where timbers have rotted, deads can suddenly collapse with disastrous consequences.

dehydrate [62]: to lose water.

dendritic [15]: of minerals: forming branch-like or mossy crystal aggregates.

density [22]: the mass of a fixed volume of a substance or rock, normally expressed in mineralogy in grams per cubic centimetre.

desktop study [87]: the examination of an area using geological maps, previous surveying/prospecting reports and peer-reviewed papers prior to the more costly business of sending a team of geologists out into the field.

development rock [98]: waste rock, both mineralized and/or unmineralized, generated when driving service tunnels and shafts in a mine.

diagenesis [35]: the conversion of a (usually buried) soft sediment into a sedimentary rock, representing all of the physical and chemical changes that occur during the process.

diamond drilling [89]: drilling of boreholes, either cored or uncored, by drills with diamond-studded drillbits that cut through the strata rapidly.

diamond pastes [73]: industrial diamonds, graded for maximum size (e.g. 1 micron, ¼ micron), suspended in a soft paste. Used where optical-grade finishes – with no scratches even under high magnification – are required, for example in reflected light microscopy.

dispersion [92]: the separation of white light into components of different wavelengths – a good example being a rainbow, where the result is a spatially arranged colour-spectrum of visible light.

displaced anomalies [54]: a geochemical anomaly not reflected anywhere in the vicinity by bedrock geology, for example alluvial gold originally transported several kilometres to an area from its source. Glacial ice is especially good at producing such anomalies, as it is such an efficient agent of both erosion and transport.

dodecahedral [27]: a polyhedral crystal with twelve regular, pentagonal faces. A crystal habit commonly adopted by members of the garnet family.

double decomposition [102]: the replacement of a more reactive metal (solid) by a less reactive metal (in solution). A well-known example is demonstrated by dropping a pinch of iron filings into a solution of copper sulphate – the result is a sludge of metallic copper, replacing the iron.

drilling-mud [91]: a lubricating medium used in diamond drilling and, in oil and gas exploration, to counter any sudden pressure-surge when reserves are tapped deep underground. To this effect, high-density additives such as finely ground baryte are added to the mix.

drusy [15]: a mineral texture in which crystals line open cavities throughout a sample of rock.

dull [21]: a mineral with little or no lustre.

dynamic metamorphism [39]: a range of processes that occur in zones of generally high tectonic strain, such as major basement-reaching faults, thrust- and shear-zones, where rock units undergo cataclasis (brittle pervasive shattering) in near-surface environments and, at depth, mylonitization caused by bulk shearing.

E

earthy [18]: a lacklustre mineral with a soil-like appearance.

economic geology [87]: the specialized branch of geology that deals with naturally-occurring commodities, often subdivided into industrial minerals, metals, gemstones and hydrocarbons.

efflorescence [63]: a bloom-like growth of secondary minerals formed on the surface of an unstable primary mineral (such as pyrite) due to its reaction with air and moisture.

electromagnetic force [1]: the force that causes interactions between electrically charged particles such as protons and electrons. Outside of gravity, it is ultimately responsible for all natural phenomena.

electron [1]: a subatomic particle carrying a single negative charge.

element [1]: a pure chemical substance made up out of unique atoms carrying a unique number of protons in their nuclei, which equates to each element's atomic number.

en cabochon [95]: a dome-shaped, smooth (i.e. non-faceted) gemstone cut.

end-members [25]: chemically pure minerals occurring at each end of a solid solution series, such as the plagioclase feldspars, with the sodium member albite ($NaAlSi_3O_8$) at one end of the series and the calcium member anorthite ($CaAl_2Si_2O_8$) at the other.

epimorph [15]: a hollow cast made by one mineral after another. Occurs when mineral X is overgrown by mineral Y but mineral X subsequently dissolves away.

epithermal [46]: of ore-deposits: low-temperature mineralization typically formed at shallow depths.

equant [7]: crystals of equal dimensions along each of their perpendicular crystallographic axes.

erosion [35]: the removal of outcropping rocks by the actions of waves, wind, rain and hail, streams and rivers, glacial ice, avalanches and frost.

euhedral [10]: perfectly crystallized.

exhalations [44]: hydrothermal fluids that have left the Earth's crust and are thus discharging in a submarine or subaerial environment.

exsolution [26]: the process through which an initially homogeneous mineral separates into two (sometimes more) different minerals. It occurs within mineral grains without the addition or removal of any compounds. Such unmixing is often a response to changes in pressure and temperature following mineral crystallization in a magma or from hot hydrothermal fluids.

extinction [76]: in petrology, when viewing sections under cross-polarized light, isotropic minerals are permanently dark (extinct) in appearance, whereas anisotropic minerals will go dark (go into extinction) four times as the microscope stage is rotated through 360°.

extraterrestrial [1]: something that has originated not on Earth but elsewhere in the Solar System or even further afield.

extrusive [32]: of igneous rocks: rocks formed from the products of volcanic eruptions at Earth's surface, in either subaerial or submarine environments.

eyepiece [74]: the part of a microscope that the user looks into. Some microscopes have one eyepiece, others have two.

F

face-centred cubic unit cell [7]: a close-packed cubic crystal lattice structure having eight identical particles on the corners of the cube and six other identical particles occupying the centres of the six faces of the cube.

faceted [94]: of gemstones: cut and polished in a specific pattern of faces in order to show off the stone's optical properties to advantage.

facies [39]: a specific set of characteristics that describes a rock and its formational environment: in mineralogy, chiefly used to classify the metamorphic rocks, each facies featuring a specific assemblage of minerals.

false floors [59]: in underground mines, a timber floor spanning a stope in order to give miners access to working faces. Covered in spoil, they are not readily apparent, and in old mines, where timbers may be rotten, there exists the risk of them collapsing under the weight of someone walking across them.

feldspars [25]: a group of aluminosilicate minerals with three pure end-members, namely potassium-feldspar (often annotated K-spar), with its two polymorphs, orthoclase and microcline, both $KAlSi_3O_8$; sodium-feldspar or albite, $NaAlSi_3O_8$ and calcium-feldspar or anorthite, $CaAl_2Si_2O_8$.

fine-fraction stream sediment [88]: sediment, collected for geochemical analysis from a stream-bed, that will pass through a very fine sieve – typically with a mesh of 0.15 mm.

flats [47]: extensive, horizontal to shallow-angled mineral deposits, typically formed as replacements of individual reactive host-rock beds such as limestones.

flaws [94]: in gemmology, internal cracks, mineral or fluid inclusions occurring within, for instance, a diamond that may mar its appearance and thereby reduce its monetary value.

fluid-inclusion studies [85]: the analysis of minute, ancient drops of hydrothermal fluids found within crystals, which trapped them as they grew. Data may be collected yielding information about fluid chemistry and the pressure and temperature conditions under which the mineralization was deposited.

free milling gold [90]: gold that is coarse enough in grain-size to be separated from the crushed ore using gravitational processes alone, such as, in the crudest method, panning.

froth flotation [87]: a process for separating valuable minerals from gangue employing the differences in their hydrophobicity, and increasing such properties through the use of surfactants and wetting agents. The ore and the reagents and water are mixed to a slurry and agitated by passing air bubbles through the mixture, the resulting – and today, very selective – separation of minerals making the processing of complex and often finely intergrown ore assemblages economically feasible.

fumaroles [46]: openings in Earth's crust, usually in volcanically active regions, which periodically to constantly emit superheated steam and a range of gases such as carbon dioxide and sulphur dioxide.

G

geobarometer [86]: a mineral indicative of certain pressure conditions at the time of its deposition.

geode [96]: an open cavity, lined with crystals.

geological processes [1]: any process involving the creation, alteration or destruction of rocks: erosion, sedimentation, diagenesis, metamorphism, melting, intrusion, eruption and deformation are common examples.

geothermal gradient [47]: the change of temperature with depth below Earth's surface.

geothermometer [86]: a mineral indicative of certain temperature conditions at the time of its deposition.

gossan [51]: the zone of iron oxides that occurs above the primary and secondary mineralized zones of a mineral deposit, formed at and close to Earth's surface where weathering tends to be at its most intense.

grade [87]: the amount of any commodity in a mineral deposit, established by various sampling procedures to ascertain whether mining is economically viable. Typically expressed in percentage terms for commoner metals and minerals and in grammes or carats per ton for rarer commodities such as gold or diamonds.

granular [18]: textural term meaning that a mineral consists of small grains or granules.

greasy [21]: of lustre: mineral has an oily or greasy appearance.

Great Oxygenation Event [50]: the transition to an oxygenated atmosphere on Earth, caused by the emergence and, during the early Proterozoic, the rapid spread of photosynthetic cyanobacteria. By 2.3 Ga, when the event had been completed, a massive dieback of anaerobic organisms had occurred – an example of a mass-extinction caused by the modification of Earth's atmosphere by a life-form.

greenhouse gas [97]: any atmospheric gas that is capable of absorbing any part of the long-wave infra-red part of the light spectrum, thereby impeding the return of some of the heat energy radiated from Earth's surface back out into space.

ground state [4]: an atom in its lowest possible energy state, with its electron(s) in their lowest energy orbitals.

H

habit [7]: the types of crystals that any mineral tends to form, e.g. tabular, acicular, etc.

hackly [19]: the ragged surface of a fracture in a native metal such as copper.

halogen [3]: the group 17 elements of the modern Periodic Table. They are all non-metals: fluorine and chlorine are gases at room temperature, bromine is a liquid and iodine a solid. The remaining member, astatine, only exists as unstable radioisotopes with very short half-lives. All halogens are highly reactive, with fluorine having the greatest reactivity: it can even form compounds with some noble gases.

hand-specimen [56]: a hand-sized chunk of rock collected either for research or, if it includes well-formed crystals, for display.

hardness [23]: a measure of how hard a mineral is: it may be relative, as in the commonly used Mohs Scale or absolute, a quantitative value.

hardness scale, absolute [23]: a quantitative scale of mineral hardness running from 1 to 400. The values are obtained by laboratory measurements that involve drawing a diamond across the sample surface under a fixed load: the width of the resultant scratch determines the value.

heavy panned concentrate [88]: the relatively dense fraction left behind when any large sample of soil, stream sediment or crushed rock is reduced, usually by panning in water, down to a standard volume, e.g. 100 millilitres. Such samples may be examined to determine what heavy minerals are present or simply sent for geochemical analysis.

hemimorphic [10]: of habit: any crystal with a different type of termination at either end.

hexagonal system [8]: a crystal system featuring four axes—three of equal lengths at 120° to one another (forming the hexagonal outline) and a fourth, perpendicular to the plane of the other three. Crystals in this system range from tabular through to long prisms or acicular forms.

hopper-crystal [10]: a crystal in which the edges are fully developed but the interiors of each face are not, although they display the same crystal symmetry so have a stepped, hopper-like appearance. This growth pattern occurs when electrical attraction is relatively intense along the edges of the crystal, so that the edges grow more quickly than the faces, where the slowest growth-rate is at face-centre.

hornfels [41]: a hard, splintery rock formed by thermal metamorphism of existing strata by intrusions of magma. The degree of recrystallization typically results in the loss of existing fabrics such as bedding.

hydraulic [45]: pertaining to the mechanical properties of any liquid.

hydrochloric acid [62]: a solution of hydrogen chloride (HCl) in water, typically prepared to a given strength. Dilute hydrochloric acid (5–10%) is used in the field to identify carbonate minerals – which effervesce when in contact with it.

hydrofluoric acid [62]: a solution of hydrogen fluoride (HF) in water. It can be used in the laboratory preparation of mineral specimens (it will dissolve quartz away from gold, for example) but it is extremely dangerous.

hydrogen bond [5]: a type of chemical bond that involves the electromagnetic interaction between polar molecules, in which hydrogen, which is electropositive, is electrically bound to an electronegative atom, such as oxygen, as in water (H_2O), which is a polar molecule with the pair of hydrogen atoms being attracted to a neighbouring oxygen atom and so on and so forth throughout.

hydrothermal [44]: literally translates as 'hot water' – in mineralogy it refers to the activity of hot underground waters that can leach metals and other elements from rocks, transport them in solution and deposit them elsewhere as minerals.

hygroscopic [62]: a hygroscopic substance is one that has the ability to attract and keep water molecules from its outside environment. Table salt (halite) is a good example, as can be readily demonstrated by leaving an open jar of salt in a room for a few weeks, especially in damp weather.

hypothermal [46]: high-temperature hydrothermal mineralization, formed at between 300 and 500 °C and often at considerable depths. Some vein-type tin-tungsten deposits associated with granite plutons fall into this category.

I, K

igneous [25]: of rocks: those that have crystallized from a melt, including both eruptive and intrusive rocks.

immiscible liquid [42]: any liquid that cannot dissolve in another – a classic example being oil in water.

inclusion [21]: a grain of one mineral wholly enclosed within another. May be formed by overgrowth of one mineral about another or by processes such as exsolution.

industrial minerals [87]: minerals of importance to industrial processes, either because they are the feedstock for an end-product or because they facilitate or catalyze other industrial processes.

inert [102]: a substance that does not tend to react with most other substances. Quartz is a good example – about the only thing that will dissolve it is hydrofluoric acid, so that it is highly persistent in natural environments where seawater is the most corrosive thing it will generally encounter.

inosilicates [26]: silicate minerals whose basic structure consists of single chains of silica tetrahedra.

interference-colours [75]: the colours exhibited by anisotropic minerals in thin section under cross-polarized light. This is because two rays of split and refracted light moving through a mineral grain will typically do so at slightly different speeds, so when they move back out into the air above a section they will be out of synchronization. The analyser will combine them into one plane of movement, but one in which there are two wavelengths interfering with one another, making the mineral look a completely different colour compared to its appearance in plane-polarized light.

intermediate rocks [34]: igneous rocks of composition midway between acidic and basic rocks, with 52–65% SiO_2, only up to 10% of it present as quartz. Darker in colour than acidic rocks, they are often pinkish-grey.

internal reflections [76]: when a translucent or transparent mineral is viewed in polished section, the light may reflect off planes within its crystals. Thus, in polished section all sphalerite appears grey, but because of its frequent translucency, reddish or sometimes yellowish internal reflections often glimmer back up at the viewer from beneath the section surface.

International Mineralogical Association (IMA) [ ]: the International Mineralogical Association is an international group of 38 national mineralogical societies, with two key aims: firstly to promote the science of mineralogy and secondly the standardization of the mineralogical nomenclature, which currently stretches to 4500 plus known mineral species, with new ones being approved by the IMA every year.

intrusive [32]: any rock that invades another, almost always in a molten state: hence intrusive igneous rocks.

ion [4]: any atom that carries a net positive or negative electrical charge, such as sodium (Na^+), copper (Cu^{2+}), oxygen (O^{2-}) or chlorine (Cl^-).

ionic bond [5]: a chemical bond formed between a metal and a non-metal: the metal gives its electron(s) to the non-metal and both metal and non-metal exist as ions within the bonded compound, their opposite charges drawing them strongly together.

iron meteorites [29]: a class of meteorites consisting mostly of intergrown nickel-iron alloys with minor sulphides and phosphides: thought to have originated from a planetoid whose interior had undergone differentiation so that a metallic core was present, prior to impact-generated breakup in the early years of the Solar System.

isotope [2]: many elements occur in nature as several isotopes in which the atomic number (number of protons) remains the same but the number of neutrons varies, so that the atomic mass numbers differ. Thus, there are three naturally occurring isotopes of carbon with the following mass numbers: 12 (6 protons and 6 neutrons), 13 (6 protons and 7 neutrons) and 14 (6 protons and 8 neutrons).

isotopic studies [85]: relative abundances of different isotopes of certain elements can be used to determine a range of properties, of which isotopic dating using carbon, lead-lead, uranium-lead, potassium-argon and other systems are the best-known: however, there are many other applications; for example, sulphur isotope abundances in sulphide minerals can help determine whether an ore deposit was deposited from fluids that had a magmatic origin.

isotropic [5]: having identical properties in all directions: in petrology this means that when viewed under crossed polars, in thin or polished section, an isotropic mineral is black in all orientations.

kimberlite [31]: a mantle-derived igneous rock that has undergone rapid transport up through Earth's crust towards the surface, assisted by volatiles such as water vapour. They commonly form pipe-shaped bodies that contain xenoliths of unmelted mantle – primarily peridotite but sometimes including diamonds, and because of the explosive nature of their upward progress they commonly incorporate fragments of the rocks they have travelled through.

L

lamellar [15]: a textural term describing aggregates of flaky, sheet-like crystals such as mica.

Late Heavy Bombardment [27]: an intense cratering event on the terrestrial planets, hypothesized to have occurred from 4.1 to 3.8 Ga. This was at a late stage of the Solar System's accretion period when the Earth and other rocky planets had formed and accreted most of their mass. The best evidence for it having happened is the cluster of radiometric dates for lunar impact melts, in samples retrieved by the Apollo missions.

lattice-point [6]: a regularly-spaced site on a crystal lattice occupied by a particle – an ion, molecule or atom – such as each corner of a cubic unit cell.

leaching [44]: the removal of elements from strata by reaction with percolating hydrothermal fluids.

lithification [47]: the conversion of a soft sediment, such as mud, to a solid rock, variably via burial-related compaction, loss of trapped water and cementation by precipitating minerals.

lithostatic pressure [45]: the pressure imposed on a rock due to the weight of all the overlying material, which will vary with burial depth and the density properties of overlying unit(s).

lode [44]: a mineral or ore-deposit filling a fissure in the rocks: interchangeable with vein.

lower mantle [30]: that part of Earth's mantle extending from approximately 660 km depth down to the core-mantle boundary at 2890 km.

lustre [21]: the way light interacts with a mineral, be it crystallized or massive: how dull or how sparkly a mineral appears. It is a non-quantitative term so can be subjective, depending on factors such as ambient light-type, the viewer's eyesight and judgement; however, it is useful when noted in conjunction with a sample's other physical properties.

M

magmatic segregation [42]: the concentration of different minerals in different parts of a magma chamber: for example, dense minerals that have crystallized early, so that they exist in a liquid, are likely to sink through the liquid to become concentrated at or near to the bottom of the chamber.

mantle [25]: that part of Earth's interior between the base of the crust (no more than a few tens of kilometres down) and its outer core, which is 2890 km down.

mantle heat-plume [32]: an upwelling of hot rock within the Earth's mantle, typically much hotter than its surroundings and able to partially melt at shallower depths, creating a so-called hotspot where magma can rise in significant quantities up towards Earth's surface, giving rise to major eruptive episodes.

massive [18]: a mineral occurrence in which no crystal structure or other texture may be discerned.

mass number [2]: the sum of the number of protons and the number of neutrons in the atomic nucleus of any isotope of an element.

mesosiderite [30]: an uncommon type of stony-iron meteorite consisting of a brecciated mixture of approximately equal amounts of metal and silicates, thought to have been impact-generated during the early years of Solar System formation.

mesothermal [46]: medium-temperature hydrothermal mineralization, typically deposited at between 200 and 300°C. Many lode-type gold deposits are of mesothermal origin.

metal loading [102]: in environmental geochemistry, the amount of various metals present in water discharges from mines.

metallic [21]: of lustre: a bright, metal-like appearance. Sulphides may weather and a freshly-broken surface is required to observe their lustre properly.

metallic bond [5]: a type of chemical bonding specific to metals, in which the outer electrons of each atom are only weakly held by the nuclei and thus readily part from them, leaving metal cations sitting in a 'soup' of electrons that move around freely. The bonding force is manifested as the multiple attractions between the nuclei and the electrons that surround them.

metamorphic – *see* metamorphism

metamorphic aureole [41]: the area around an igneous intrusion in which the host-rocks show changes caused by contact-metamorphism.

metamorphism [25, 38]: changes to a rock's original mineralogy and texture, brought about by heat and/or pressure acting in various degrees of severity.

metasediments [39]: rocks that were clearly once sediments but whose mineralogy and textures demonstrate that they have undergone changes due to heat and/or pressure.

metasomatism [47]: the replacement of reactive rocks such as limestones by other minerals due to interactions with, for example, hydrothermal fluids.

meteoric waters [47]: groundwaters that have been almost entirely derived from rainfall.

meteorites [25, 27]: fragments of rock from other parts of the Solar System that have been captured by Earth's gravitational pull and have made it down to the surface without being entirely vaporized.

micromounting [56]: the study of small to microscopic crystals of various minerals on typically small (0.5–2 cm) samples mounted in clear plastic boxes to protect them from dust.

micron [22]: a measurement unit typically used in petrography when describing minerals present in thin and polished sections under high magnification: 1000 microns = 1 millimetre.

Miller indices [11]: a precise mathematical way to describe crystals; it involves assigning a unique numerical value to each crystal face or cleavage plane, defined by determining how they intersect the main crystallographic axes.

mineral assemblage [39]: any group of minerals occurring together.

mineral deposit [25]: any atypical concentration of minerals in comparison to their normal crustal occurrences: e.g., quartz is common in granite but a five-metre thick vein of solid quartz is less typical when one considers the crust as a whole.

mineral species [7]: a mineral species is a unique element or chemical compound (or structural polymorph of same) that is normally crystalline and that has been formed as a result of geological processes.

minerals industry [87]: generally regarded as the inorganic sector of the extractive industries; it includes both quarrying and mining and it produces metals, industrial minerals and stone for various purposes.

mobile [102]: any chemical whose properties, for example solubility, mean that it spreads through the natural environment with relative ease.

Mohs scale [23]: relative scale of hardness used in fieldwork as a mineral identification aid and relying on the usage of common objects such as copper coins, penknife blades, etc.

monoclinic [8]: a crystal system that has three axes of varying length, two of which meet at an oblique angle and a third that is perpendicular to the other two. Crystals have a single 2-fold main axis of rotational symmetry; most monoclinic minerals occur in prismatic habits of varying length.

mudstones [35]: sedimentary rocks originally deposited as muds in lakes, sluggish rivers, marine basins and other low-energy environments.

mylonite [41]: a rock produced by dynamic metamorphism at depth in Earth's crust and at high temperatures and pressures. Under such conditions, rocks become relatively ductile and may deform by shearing in bulk, creating a metamorphic rock with a streaky appearance.

N

neutron [1]: a subatomic particle that has no net electric charge and a mass slightly larger than that of a proton, together with which they make up the nuclei of atoms.

nickel laterites [52]: upgraded, often economically workable nickel deposits formed by the intensive weathering of olivine-rich ultrabasic intrusive rocks such as peridotite, typically under tropical conditions, and leading to the re-precipitation of nickel as oxides and silicates with a concentration increase that may be tenfold.

noble gases [3]: the group 18 elements in the modern periodic table, and in all cases comprising odourless and colourless gases with extremely low chemical reactivity due to having their outer electron shells filled. The six noble gases that occur naturally are helium (He), neon (Ne), argon (Ar), krypton (Kr), xenon (Xe), and the radioactive and locally hazardous radon (Rn).

nucleus [1]: the extremely dense inner part of an atom, consisting of protons and neutrons.

O

objective lenses [74]: the high-powered magnifying lenses situated at the bottom of a microscope, closest to the sample. In petrological microscopes there may be three or more objectives of different magnifying powers that may be rotated into position.

oceanic crust [34]: the relatively thin (7–10 km) crust of basic igneous rocks, topped with sediments of varying thicknesses, that underlies Earth's deep oceans. It is generated at mid-oceanic ridges but is relatively short-lived compared to continental crust because it is readily subducted at destructive plate margins.

ochre [100]: generic term for colourful hydrated oxides of iron, typically found in reddish through to yellowish shades.

olivine [26]: a group of orthorhombic rock-forming silicates, named after their greenish colour and consisting of a solid solution series between magnesium (forsterite, Mg_2SiO_4) and iron (fayalite, Fe_2SiO_4) end-members.

opaque [22]: a substance through which light will not pass, even at thin section thickness.

ore [87]: in general, an ore is a mineral from which a metal is obtained; for example, galena is an ore of lead.

ore-deposit [22]: specifically, in economic geology, an ore-deposit is one that is economically viable to work, although the viability may change through time as metal prices fluctuate. However, in more general geology, ore-deposits are simply concentrations of ore-minerals, whether mined or not.

orefield [87]: a district from which particular metals have been mined or are being mined from multiple localities.

orthorhombic [8]: a crystal system that has three mutually perpendicular axes, but in which each axis has a varying length. There are three main 2-fold axes of rotational symmetry. Orthorhombic minerals typically form tabular to prismatic crystals.

oversaturation [42]: an over-abundance of an element in a melt; for example, if a magma assimilates a sulphide-rich black shale, it will almost certainly become oversaturated in sulphur.

oxalic acid [62]: an organic compound with the formula $H_2C_2O_4$, and obtained as white crystals. The crystals are dissolved in water to make a solution that will dissolve the iron-staining that often coats quartz crystals when they are found in the field. It must be used with care, however; firstly it is toxic, and secondly, it can etch some other minerals.

oxidation [4]: the process of electron-removal to form cations. For example, pyrite is iron(II) sulphide, FeS_2, and it often oxidizes to goethite, $FeO(OH)$, iron(III) oxide-hydroxide. Oxidation processes like this are important in mineral weathering.

P, Q

palaeoplacer [53]: an ancient deposit of alluvial origin, now a sedimentary rock such as a sandstone or a conglomerate that carries a heavy mineral suite with minerals such as cassiterite or gold being present.

pallasite [30]: an uncommon type of stony iron meteorite consisting of large (centimetre-scale) crystals of olivine set in a matrix of iron-nickel alloys.

paragenesis, paragenetic sequence [73]: the order of coming into being of a suite of minerals in an igneous or metamorphic rock or a hydrothermal mineral deposit, determined through petrological microscopy involving the detailed recording of mineral relationships in multiple thin or polished sections.

paragenetic diagram [84]: diagrammatic way in which to illustrate the paragenetic sequence of a mineral assemblage, usually presented to complement a written description of the minerals and their relationships.

partial melting [32]: localized melting of some or most of a rock that occurs in situations that favour the process, such as in the depths of subduction zones where seawater, subducted with oceanic crust, lowers the melting point of that crust and the crust in contact with it. Partial melts that liquidize the less refractory constituents of a rock, whilst leaving more refractory components unmelted, produce magmas with a different composition to the bulk composition of the melted crust. For example, in a subduction zone where descending basaltic oceanic crust and overlying granitic continental crust both undergo some melting, a magma of intermediate composition will be generated.

pathfinder elements [88]: elements whose elevated levels in the environment suggest the presence of potentially economic mineral deposits: for example, arsenopyrite often accompanies gold deposits so that arsenic anomalies in stream sediments or soils would flag up a district as potentially auriferous.

pearly [21]: of lustre: a pearl-like sheen.

pegmatite [33]: a very coarse-grained igneous rock that is typically granitic, but which may be of intermediate or basic composition depending on the type of igneous activity in the area. Often forms dykes or vein-like structures, and some pegmatites feature large cavities lined with euhedral crystals.

perfect cleavage [18]: cleaving of a crystal that leaves smooth, planar surfaces.

Periodic Table [3]: the tabular arrangement of the chemical elements, arranged into groups by their atomic numbers, electron configurations and chemical properties.

perthitic texture [26]: an intergrowth of two feldspars involving a host (potassium-rich alkali feldspar such as orthoclase) carrying exsolved lamellae of sodic feldspar such as albitc.

petrography [73]: the detailed description of a mineral assemblage: a branch of petrology.

petrographical microscope [22]: a microscope designed for the optical study of mineral assemblages at high magnification in thin or polished sections of samples taken from rocks or mineral deposits.

petrology [73]: the branch of geology that studies the origin, composition and internal structure of rocks.

pH [14]: the measure of the acidity or basicity of an aqueous solution: pure water is close to neutral (pH7) and solutions with a higher value (up to 14) are said to be increasingly alkaline, while solutions with a lower value (down to 0) are said to be increasingly acidic.

phaneritic [32]: a textural term pertaining to igneous rocks and meaning that the minerals making up the rock's matrix are all sufficiently coarse-grained to be seen with the naked eye.

phenocryst [33]: in igneous rocks with a porphyritic texture, a large crystal set in a finer groundmass.

phosphorites [93]: sedimentary rocks containing at least 15–20% phosphate minerals, which is a significant enrichment of phosphorus.

pinacoids [13]: in crystallography, crystal faces that cut the vertical c axis and are parallel to the horizontal a1 and a2 axes, resulting in a pair of parallel faces.

placer mineral deposits [53]: alluvial or beach sediments in which river or wave action has sorted mineral grains according to their density, resulting in enhanced concentrations of heavy minerals such as gold.

polarizer, plane-polarized [75]: normally, light-waves vibrate outwards at right-angles to their overall path. Passing a ray of light through a polarizer blocks all but those vibrating in one orientation, producing plane-polarized light.

planetary differentiation [27]: the process by which, in the molten state, the interior of a planet is sorted according to the relative densities of its constituents by gravity. Hence the densest material (iron and nickel) ends up in the core and the lightest material (aluminosilicates) tends to make up the crust. This is thought to occur early in the life-cycle of a planet.

planning stage [90]: in mineral exploration, the point at which it has been decided that a mineral deposit is economically feasible to extract: the company must then go to the relevant planning authority to seek approval for its proposals.

plate tectonics [29]: the scientific theory that describes the large-scale motion of Earth's lithosphere, consisting of plates of continental and oceanic crust, underpinned by the uppermost part of the mantle, developed during the early part of the latter half of the twentieth century.

pleochroism [76]: the optical phenomenon in which an anisotropic substance or mineral, viewed under a petrographical microscope, exhibits different colours when observed at different angles.

polished section [22]: a chip, typically 1–2 cm across, sampled from a rock or mineral deposit, embedded in hard-setting resin, ground flat and then polished with fine-grained abrasive pastes to an optical-grade finish for study at high magnification using a reflected light microscope or scanning electron microscope.

polished thin section [73]: a standard thin section with a polished upper face, ideal for readily examining both opaque and non-opaque mineral phases using a petrographical microscope that may be switched from transmitted to reflected light mode and back.

pollution [97]: the presence in or introduction into the environment of a substance that has harmful or poisonous effects.

polymetallic [102]: of mineral deposits: those containing a large number of metals (as opposed to, for example, just lead and zinc).

polymorphs [14]: minerals that have an identical chemical composition but which crystallize in different systems, for example diamond (cubic) and graphite (hexagonal).

pore-water [38]: groundwater held in a sedimentary rock in the pores (spaces) in between grains.

porphyritic [33]: an igneous rock-texture in which large crystals (phenocrysts) of one or more minerals occur embedded in a finer-grained groundmass.

porphyroblasts [41]: phenocryst-like crystals occurring in metamorphic rocks.

porphyry-type mineralization [46]: a class of mineral deposit associated with acidic or intermediate intrusive igneous rocks, in which the ore minerals occur throughout the rock as fine disseminations typically situated along microfractures caused by bulk boiling of hydrothermal fluids. An important source globally of metals like copper, molybdenum, gold and tungsten.

powder colour [21]: the colour of a mineral when finely powdered. In the field, this can be determined by drawing a fragment of a mineral across a rough plate of unglazed white porcelain, producing a streak.

precious stones [93]: traditional classification term for gemstones: diamond,

ruby, sapphire and emerald are regarded as precious, with the rest being termed semi-precious.

pressure-fringe [82]: A growth of minerals, typically quartz, calcite, chlorite or mica, forming a zone alongside the face of a crystal (such as a pyrite cube) in a rock modified during regional metamorphism involving tectonic strain, such as slate. During strain-related metamorphism, minerals are locally dissolved from the zone of high pressure, where the matrix is compressed against the crystal. They are redeposited in the relatively low pressure zone at the side of the crystal that is normal to the maximum compressive stress.

primary minerals [51]: in mineral deposits, the minerals that were deposited by the original mineralizing process.

prismatic [8]: a crystal habit: elongated crystals with parallel faces and of similar thickness in cross section anywhere along their length.

prism faces [13]: the four sides of a crystal that are parallel to its c-axis, which are short if the crystal is tabular or long if it is prismatic.

proton [1]: a subatomic particle carrying a single positive charge, making up, with neutrons, an atom's nucleus.

provenance [35]: the origin or source area of particles within a rock. For example, the populations of tiny zircons within sandstones are often radiometrically dated by the U-Pb method in an attempt to determine their original source terrane(s).

pseudomorph [14]: a mineral replacing a crystal of another mineral, creating a partial to complete replica of the original crystal.

pyrite-decay [63]: the process by which a specimen containing pyrite can be destroyed due to the reaction between the pyrite and air and moisture, which in turn releases sulphuric acid that attacks other minerals in the sample.

pyritization [45]: a common type of wall-rock alteration associated with mineral deposits, in which certain minerals become replaced by pyrite. Can also refer to the replacement of fossils by pyrite.

pyritized [38]: replaced by pyrite.

pyroxenes [26]: a group of dark-coloured rock-forming silicate minerals found in many igneous and metamorphic rocks. They share a common structure, consisting of single chains of silica tetrahedral, and they crystallize in the monoclinic (clinopyroxene) and orthorhombic (orthopyroxene) systems. Pyroxenes have the following general formula: $XY[(Si,Al)_2O_6]$ in which X is typically calcium, sodium, ferrous iron and magnesium and Y is typically chromium, aluminium, ferric iron, magnesium and manganese.

quenching [32]: the rapid cooling of magma, usually caused by its contact with water or air during eruption. Volcanic glass is formed by quenching.

R

reduction [4]: any chemical reaction that involves the process of electron-gain. For example, the reduction of a mineral containing copper in its 2+ oxidation state leads via its 1+ oxidation state to the native metal.

reflectance, reflectivity [76]: the percentage of ambient light that is reflected back by a mineral in a polished section: a measure of how bright a mineral appears when viewed under a reflected light petrographical microscope.

reflected light microscopy [22]: a branch of petrographical microscopy that deals with opaque minerals, which are viewed in the light that is reflected from their surfaces in polished sections.

refractory [92]: of material science: a substance that maintains its structure at high temperatures. In mineralogy, zircon is a good example: when a rock undergoes partial melting, any zircons present may well survive intact.

regional metamorphism [39]: metamorphism affecting the rocks over a wide, as opposed to localized, area.

relief [78]: in thin sections, relief refers to minerals that stand out from the background because their refractive indices have a marked difference with that of the resin used to mount the rock to the slide. In polished sections relief is a topographical property caused by softer minerals being more easily abraded than harder ones during the polishing process. Polished sections with high relief are useless because it obscures many important mineral relationships.

remote-sensing [87]: in mineral exploration, examining a district without actually going into the field. Typically this will involve poring over satellite imagery to look for any features of interest such as crustal lineaments. This may be useful when deciding where exploration ought to be concentrated, although it is no substitute for detailed fieldwork!

reniform [15]: a textural term meaning kidney-shaped masses, as in the well-known variety of haematite, 'kidney-ore'.

rescue-collecting [61]: the collecting of large amounts of specimens from an occurrence which is about to be rendered inaccessible or destroyed, for example by road-building or quarrying.

resinous [21]: of lustre: having a resin-like appearance.

reticulated [15]: a textural term that refers to needle-like crystals occurring in a crystographically oriented criss-cross pattern.

retrograde metamorphism [40]: the recrystallization of high-grade metamorphic rocks to lower-grade rocks.

rock-forming minerals [25]: common minerals that make up the bulk of igneous, sedimentary or metamorphic rocks, such as quartz or feldspar.

rosette [18]: a textural term that refers to platy crystals forming spherical aggregates with the appearance of a rose-flower.

rotational symmetry [7]: the property of an object (in this context a crystal) that looks the same after a certain amount of rotation. For example, a cubic crystal lying flat on a desktop looks the same when rotated through 90°: the process may be repeated three more times to the same effect, so that it has a four-fold axis of rotational symmetry through the centre of each face.

S

sandstones [35]: sedimentary rocks predominantly composed of sand-sized grains, 0.0625 to 2 mm in diameter.

scanning electron microscope [78]: a powerfully magnifying microscope that produces images of samples using a focused beam of high-energy electrons.

secondary electrons [79]: in scanning electron microscopy, electrons that are ejected by the specimen due to interaction with the electron beam.

secondary minerals [51]: in mineral deposits, any minerals that have formed through processes occurring after the mineralizing event, such as weathering.

sedimentary [25]: in rock classification, sedimentary rocks are those whose origin involved deposition of sediments by water, wind, ice and mass-movement, plus chemical precipitates such as limestone and evaporites formed by the evaporation of bodies of saline water.

sedimentary basin [47]: a depressed or subsided area of the Earth's crust where sediments are accumulating in either a marine or terrestrial environment.

sedimentary-exhalative [47]: a class of mineral deposit in which hydrothermal fluids have been expelled onto a sea-floor: over lower or depressed areas of the sea-bed the relatively heavy, hot brines collect and are mixed with cool seawater. The mixing reduces metal and sulphur solubility, so that metal sulphides precipitate out and are deposited as sedimentary layers. Prolonged and undisturbed activity of this type can allow significant amounts of sulphides and other minerals such as baryte to accumulate.

semi-precious stones [93]: traditional classification term for gemstones: all other stones apart from diamond, ruby, sapphire and emerald.

septarian nodule [38]: sedimentary concretions with a distinctive internal structure consisting of a network of cracks (septa) dividing them up internally into sections.

sericitization [45]: a common type of wall-rock alteration associated with mineral deposits, in which certain minerals become replaced by the white mica, sericite.

series [25]: a compositional spectrum of minerals between two pure end-members, such as the sodium- and calcium-bearing plagioclase feldspars that form the series between albite and anorthite, the pure sodium and calcium end-members respectively.

shell [25]: in atoms, the series of electron orbitals surrounding the nucleus, each being a zone of stable energy levels.

shock metamorphism [39]: a relatively infrequent process that occurs when a sizeable meteorite or asteroid collides with the Earth's surface, a process typically involving instant heating to high temperatures and the imposition of massive pressures.

silicification [45]: a common type of wall-rock alteration associated with mineral deposits, in which certain minerals become replaced by silica.

silky [21]: of lustre: silky in appearance.

siltstones [35]: sedimentary rocks predominantly composed of silt-sized grains, 0.0625 mm or less in size and barely visible to the naked eye.

simple cubic unit cell [6]: a basic

building-block of a mineral that crystallizes in the cubic crystal system. Each corner of the cube, known as a lattice-point, is marked by a particle – an atom, ion, or molecule – depending on the mineral species. The edges of the unit cell all connect identical particles. Because a cube has eight corners, a minimum of eight identical particles has to be present to make a simple cubic unit cell.

Sites of Special Scientific Interest (SSIs) [58]: in Earth Sciences, localities where one or more aspects of their geology are deemed to be of national or international importance to science, and are thereby protected by law.

skarns [50]: calcium-bearing silicate-rich rocks, typically formed by metasomatism at the contact zone between intrusive rocks and carbonate rocks such as limestones. Depending on the nature of the intrusion and the associated hydrothermal activity, they may contain a wide variety of minerals.

sluice [54]: a trough-shaped wooden, plastic or steel box, with part of its floor featuring a series of riffles – bars that cross from one side to the other. This is lain in a river-bed so that water flows through it at a steady rate; river sediments are fed through and any heavier minerals are trapped behind the riffles because of the way that they perturb the current. Set up properly, a sluice can process large volumes of sediment in a relatively short timeframe, compared to panning.

smelting [54]: the process of reducing metalliferous minerals in order to produce their constituent metals. In its simplest form, this involves roasting an ore to convert it to an oxide and then heating with charcoal to remove the oxygen. Copper, lead and tin were prepared in this simple manner in ancient history; however, with some metals the process is a lot more complicated.

smelting flux [91]: a substance that, when added to a mixture to be smelted, allows the non-metalliferous slag to melt at a lower temperature, thereby making it more runny and easy to draw off.

Snowball Earth [51]: a past period during which the Earth was almost completely covered by ice; several major glaciations of this nature occurred during the Cryogenian period of the Late Proterozoic, 850–635 million years ago.

solid solution series – *see* **series**

specific gravity [22]: a numerical value that expresses the density of a mineral compared to that of water, calculated by dividing the mass of the sample by the measured volume of water it displaces when placed in a graduated vessel such as a beaker.

splintery [19]: of fracture: a mineral that breaks up into lots of sharp little splinters.

stability field [14]: the range of physical and chemical conditions within which any mineral is stable.

stage [75]: the rotatable part of a petrographical microscope upon which the thin or polished section is mounted for examination.

stalactitic [15]: of texture: an elongated aggregate of crystals that has a smooth surface.

stellate [15]: of texture: multiple crystals all radiating outwards from a common centre, like a cartoon star.

stockwork [44]: in mineral deposits, numerous thin veins or veinlets of minerals, occurring throughout a large body of rock.

stony meteorites [29]: meteorites that consist mainly of silicate minerals, and subdivided into the chondrites and achondrites.

stopes [59]: in underground mining, areas – often of great extent – where mineral veins have been mined away. They are often prone to collapse.

strain-related twinning [14]: the formation of twinned crystals in pre-existing minerals due to tectonic strain.

streak [21]: the colour of a mineral when finely powdered. In the field, this can be determined by drawing a fragment of a mineral across a rough plate of unglazed white porcelain, known as a streak-plate.

strewnfield [29]: many large meteors explode during their descent through Earth's atmosphere: the strewnfield is the area in which the resulting fragments have rained down onto the ground.

Strunz-Nickel Grouping [24]: a scheme for categorizing minerals based upon their chemical composition. It was introduced in 1941 and revised versions of it have been used ever since, the most recent revision being the ninth edition published in 2001, though a tenth edition is pending.

sub-alkaline [34]: basalts erupted at mid-ocean ridges have a relatively low alkali metals content (compared, for example, to the alkali basalts erupted at island arcs) and are referred to as sub-alkaline or tholeiitic.

subduction [32]: the process whereby one tectonic plate (usually oceanic crust) is forced beneath another at a destructive plate margin, diving down into the mantle where partial melting occurs.

subhedral [10]: an imperfect crystal with some, but not all, faces developed and recognizable.

sub-metallic [21]: of lustre: somewhat metallic in appearance.

sutured [82]: mineral grain boundaries with an interlocked, wavy outline. Diagnostic of quartz that has recrystallized under tectonic strain.

symplectite [82]: an unmixing or exsolution texture that resembles drops of oil in water but which is very fine-grained, requiring high magnification to observe it in any detail.

systematic collectors [68]: mineral collectors who collect just one class of minerals, such as carbonates, arsenides or native metals.

T

tabular [8]: a crystal habit in which the mineral forms flat plates.

tailings [90]: the waste left over from the milling operation at a mine, consisting (depending on the recovery process used) of sand- to clay-sized articles of gangue and wall-rock, but often with some residual metallic minerals present, which may lead to pollution problems if they get into solution.

tetragonal [8]: a crystal system that has three mutually perpendicular axes; two axes are of equal length, while the third vertical axis is of varying length and can be either shorter or longer than the other two. There is a main 4-fold rotational symmetry axis with up to four 2-fold axes; tetragonal minerals can form tabular (platy), equant or prismatic habits.

tetrahedra [10]: crystals with four triangular faces.

thin section [21]: a 30 micron thick slice of rock, prepared by impregnation with resin (to stop fracturing), followed by careful cutting and grinding to the correct thickness, and intended for transmitted light microscopy.

tholeiite [34]: a basalt erupted at a mid-ocean ridge, which characteristically has a relatively low alkali metals content (compared, for example, to the alkali basalts erupted at island arcs).

tonnage [87]: the amount, in tonnes, of ore assessed to be present in an ore-deposit.

trace-element geochemistry [86]: in research: the determination of levels of certain rare elements present in minute amounts in common minerals using high-sensitivity techniques. Such data may give information as to the origin and/or affinities of a rock unit or mineral deposit.

trace-elements [21]: elements that occur in minute quantities within the crystal lattice of a mineral, having had their ions substituted for those of the main elements, a process made possible by similar particle sizes.

transition metal [3]: the elements occupying columns 3–12 of the periodic table. They have varying numbers of valence electrons because they are generally able to make available electrons from the next shell down as well as the valence-shell. This property means that they can occur in a range of oxidation-states.

translucent [21]: a mineral through which light can pass.

transparent [21]: a mineral that one can see through, just like clear glass.

triclinic [8]: a crystal system that has unequal axes, all meeting at oblique angles–in other words, the crystals are asymmetric. Crystals of triclinic minerals tend to be squat or tabular in nature.

trigonal [8]: a subdivision of the hexagonal crystal system in which the single main axis of rotational symmetry is 3-fold as opposed to 6-fold.

twinning [13]: a phenomenon that occurs when two separate developing crystals of a mineral share some of the same crystal lattice points in specific, symmetrical arrangements known as twin laws.

U

ultrabasic [34]: rocks that contain less than 45% SiO_2 and consist essentially of ferromagnesian minerals such as pyroxene, olivine and spinel (chromite and/or magnetite), with up to 10% of plagioclase feldspar.

unit cell [6]: an orderly pattern of lattice-units that repeats itself three-dimensionally to the edges of a crystal.

upper mantle [30]: that part of Earth's mantle that starts at the base of the crust and extends down to 410 km below the surface.

V

valence [3]: the number of chemical bonds that an atom can form with other adjacent atoms, controlled by the number of electrons and empty spaces present in the element's valence shell.

visible spectrum [75]: that part of the light spectrum that can be seen by the human eye, from violet (wavelength 390 nanometres) through to red (wavelength 760 nanometres).

vitreous [21]: of lustre: glassy in appearance.

Volcanogenic Massive Sulphide (VMS) [46]: a type of metal sulphide ore deposit, of submarine exhalative character, but which forms as a result of hydrothermal activity directly associated with volcanism.

vug [10]: a cavity in a mineral deposit or rock that is lined with euhedral crystals.

W

wall-rock alteration [44]: the chemical and mineralogical alteration of rocks through which hydrothermal fluids have passed, usually connected with the formation of mineral deposits. Its intensity and extent depends on several variables: for example, the porosity of the rock, its mineralogy and the chemistry of the fluids.

waxy [21]: of lustre: a wax-like appearance.

weathering agents [1]: substances that can attack minerals at or close to Earth's surface, such as carbon dioxide, which when dissolved in rainwater forms a weakly acidic solution that can slowly dissolve many silicates.

Widmanstätten pattern [29]: the intergrowth of crystals of nickel-iron alloys, typically taenite and kamacite, which form iron meteorites. On a polished surface, the pattern can be revealed by etching with acid, because taenite is more etch-resistant.

winzes [59]: shafts connecting underground sections of mines.

X

xenocryst [31]: an exotic crystal captured and transported by a magma and thus eventually incorporated into an igneous rock.

xenolith [31]: an exotic rock-fragment captured and transported by a magma and thus eventually incorporated into an igneous rock. Many intrusions contain xenolith-rich zones.

X-ray diffraction (XRD) [65]: an analytical technique that is used to determine a mineral's identity via its detailed and unique crystal structure. When bombarded with X-rays, any crystalline mineral will produce its own unique pattern of reflected X-rays that form a series of well-defined peaks that the technique detects and measures.

X-rays [65]: a form of radiation with very short wavelengths varying from 0.1 to 10 nanometres.

Further reading and resources

Books:

Minerals of Britain and Ireland, A.G. Tindle, Terra Publishing, Harpenden (now Dunedin Academic Press), 2008. ISBN: 9781903544228

Manual of the mineralogy of Great Britain & Ireland, R. P. Greg and W. G. Lettsom, John van Voorst, London, 1858.

The Ore Minerals and their Intergrowths, P. Ramdohr, Pergamon, Oxford (now Elsevier), 2nd Revised edition, 1980. ISBN: 9780080116358

Atlas of Opaque and Ore Minerals in Their Associations, R.A. Ixer, Van Nostrand Reinhold (now John Wiley), 1992. ISBN: 9780442302917

Atlas of the Rock-Forming Minerals in Thin Section, W.S. Mackenzie, Routledge, London, 1980. ISBN: 9780582455917

Atlas of Igneous Rocks and Their Textures, W.S. Mackenzie, C.H. Donaldson and C. Guilford, Prentice Hall, New Jersey, 1982. ISBN: 9780582300828

Atlas of Sedimentary Rocks Under the Microscope, by A.E. Adams, W.S. Mackenzie and C. Guilford, Routledge, London, 1984. ISBN: 9780582301184

An Introduction to Metamorphic Petrology, B.W. Yardley, Prentice Hall, New Jersey, 1989. ISBN13: 9780582300965

Igneous Rocks and Processes: A Practical Handbook, R.Gill, Wiley-Blackwell, Chichester, 2010. ISBN: 978-0632063772

Optical Mineralogy: Principles and Practice, C.D. Gribble and A.J. Hall, CRC Press, Florida, 1993. ISBN: 9781857280142

Manual of Mineralogy, J.D. Dana, Merchant Books (now Wiley), 2008. ISBN: 9781603861021

Introduction to Mineral Exploration, C. Moon (ed), M. Whately and A.M. Evans, Wiley, Chichester, 2nd edition, 2005. ISBN: 9781405113175

Practical Gemmology, D. Cunningham (revised; original by R. Webster), NAG Press, London, 2011. ISBN: 9780719804311

An Introduction to Environmental Chemistry. J.E. Andrew, P. Brimblecombe, T.D. Jickells, P.S. Liss, B. Reid, Wiley-Blackwell, Chichester, 2nd Edition 2003. ISBN: 978-0632059058

Websites:

www.mindat.org (accessed 31.10.2014)

www.minerals.net (accessed 31.10.14)

www. webmineral.com (accessed 31.10.2014)